Brain's Hidden Switches: Beyond DNA

Veron

Table of Contents

Contents

Abbreviations

ACN	Acetonitrile
AD	Alzheimer's disease
ALS	Amyotrophic lateral sclerosis
APP	Amyloid-precursor protein
Aβ	Amyloid-β
BH	Benjamini-Hochberg correction
CA1	Cornu ammonis 1
CERAD	Consortium to Establish a Registry for Alzheimer's Disease
Cont.	Control
cDNA	Complementary DNA
ddH$_2$O	Double distilled water
DAPI	4',6-Diamidino-2-phenylindole (double stranded DNA staining)
DIA	Data independent acquisition
DMSO	Dimethyl sulfoxide
DEPs	Differentially expressed proteins
DNA	Deoxyribonucleic acid
DTT	Dithiothreitol
EC	Entorhinal cortex
EDTA	Ethylenediaminetetraacetic acid
EOAD	Early onset AD
ESI	Electrospray ionization
FA	Formic acid
FAD	Familial Alzheimer's disease
FBS	Fetal bovine serum
FC	Fold change
FDR	False discovery rate
FTD	Fronto-temporal dementia
GAPDH	Glyceraldehyde 3-phosphate dehydrogenase
GSK3-β	Glycogen synthase kinase 3-β
GO	Gene Ontology
HD	Huntington's disease
HMW	High molecular weight
hr	Hour
IAA	Iodoacetamide
IB	Immunoblotting
IF	Immunofluorescence
IgG	Immunoglobulin G
IHC	Immunohistochemistry
Iβ-1	Importin-β-1
kDa	Kilodalton
LFQ-MS	Label-free quantification-mass spectrometry

LB	Lauria-Bertani
LMW	Low molecular weight
LOAD	Late onset Alzheimer's disease
MAPT	Microtubule-associated protein tau
MCI	Mild cognitive impairment
MMSE	Mini-Mental State Examination
MS	Mass spectrometry
NFTs	Neurofibrillary tangles
NP-40	Nonidet P-40
PBS	Phosphate-buffered saline
PBS-T	Phosphate-buffered saline-Tween-20
PS	Penicillin-Streptomycin
PD	Parkinson's disease
PLAAC	Prion-like amino acid composition
PLD	Prion-like domain
PRNP	Prion protein gene
PSEN1	Presenilin 1
PVDF	Polyvinylidene difluoride
p-tau	Phosphorylated tau
qRT-PCR	Quantitative real time-polymerase chain reaction
Q-TOF	Quadrupole-time-of-flight
RBD	RNA-binding domain
RBP	RNA-binding protein
RNA	Ribonucleic acid
Rnq1	[PIN+] prion protein
RNP	Ribonucleoprotein
ROS	Reactive oxygen species
rP	Pearson's linear correlation coefficient
rpAD	Rapidly progressive Alzheimer's disease
rpm	Revolutions per minute
RT	Room temperature
SAD	Sporadic Alzheimer's disease
sCJD	Sporadic Creutzfeldt-Jakob disease
SDS	Sodium dodecyl sulfate
SDS-PAGE	Sodium dodecyl sulfate-polyacrylamide gel electrophoresis
spAD	Sporadic Alzheimer's disease
SG	Stress granule
SFPQ	Splicing factor proline and glutamine rich
SWATH-MS	Sequential window acquisition of all theoretical mass spectra-Mass spectrometry
Sup-35	Eukaryotic peptide chain release factor GTP-binding subunit
Tau	Tubulin-associated unit
TBS	Tris-buffered saline

TBS-T	Tris-buffered saline-Tween-20
TEMED	Tetramethylethylenediamine
TFA	Triflouroacetic acid
TIA-1	T-cell intracellular antigen-1
tM	Threshold Mander's coefficient
Tris	Tris(hydroxymethyl)aminomethane
UTRs	Untranslated regions
VCP	Valosin-containing protein

Summary

Alzheimer's disease (AD) is the most prevalent cause of dementia. Typically, AD is characterized as a slow progressive dementia with an average disease duration of eight years. Classically, AD is categorized into two subtypes: the first subtype includes cases with spontaneous onset, termed sporadic AD (spAD), while the second subtype (familial AD) includes cases exhibiting mutations in genes encoding presenilin-1, presenilin-2, and amyloid-precursor protein (APP). Recently, a rapidly progressive variant of Alzheimer's disease (rpAD) was identified, in which patients exhibit a rapid cognitive decline and/or short disease duration (average of 4 years). It is known that spAD and rpAD share core neuropathological features, but unfortunately the altered molecular processes, which eventually lead to these variable rates of progression, remain elusive. To this end, we aimed to explore the emerging role of RNA-binding proteins (RBPs) in these two AD subtypes and in sporadic Creutzfeldt-Jakob disease (sCJD), another rapidly progressive form of dementia.

In the current study, we utilized an RNA pull-down approach from brain samples followed by mass spectrometry analysis to comprehensively interrogate RNA-binding protein (RBP) complexes; these were examined in human brain frontal cortex samples from three groups of patients, namely spAD, rpAD, and sCJD, as well as controls. Using a combination of bioinformatic and computational techniques, significant targets from the proteomic study were identified and prioritized for further characterization. The first set of analyses investigated differential expression of a target RBP termed splicing factor proline and glutamine rich (SFPQ) at both the protein and mRNA level, its accumulation as well as its possible interactions with tau protein and stress granules (SGs) in the postmortem brains. To investigate a mechanistic link between SFPQ and the pathogenesis and progression of AD, it was furthermore studied in two cellular models – the cellular model of stress and the tau-pathology model – given that in the human brain associations exist between SFPQ, stress granules and tau protein. Finally, SFPQ and associated proteomic signatures were studied at the pre-symptomatic and symptomatic stages of the disease in the 3xTg-AD mice model, in order to uncover very early changes occurring during the disease progression.

Summary

In the present study, the RNA-binding proteome from Alzheimer's and sCJD sub-types were identified and characterized. The proteomic investigation, in combination with several bioinformatic and computational approaches, highlighted quantitative and qualitative changes in the identified RNA-binding proteome in a disease-subtype-specific manner. We identified a dysregulation pattern both at the protein and mRNA level, including the dislocation of the RNA-binding protein SFPQ as a novel pathological target in the rapidly progressive subtype of AD. The SFPQ protein is involved in multiple functions in the brain, including splicing, transcription, and transport of mRNAs. This suggests that the dysregulation/dislocation of SFPQ lead to defects in these functions, which aggravate the neurodegenerative processes and eventually contribute to the rapid progression.

Furthermore, co-immunofluorescence analysis revealed a change in the fluores-cence pattern of phosphorylated tau (p-tau) along with SFPQ, with a complete nu-clear depletion of both proteins and co-localization in the perinuclear/cytoplasmic area. This indicated that there are changes in the function of both nuclear tau and SFPQ. The cytoplasmic SFPQ showed co-localization with TIA-1, a marker of stress granules (SGs). In parallel with human brain findings, our study of the cellular model of stress indicated that SFPQ and tau translocate into the cytoplasm to form SGs after oxidative stress treatment. This translocation of the two proteins into the SGs provides a possible mechanism for the observed depletion/dislocation of SFPQ and nuclear tau in postmortem human brains from rpAD cases. At initial phases of the disease, kinases phosphorylate not only tau but also SFPQ, leading to their translo-cation into perinuclear/cytoplasmic area and their incorporation into stress granules. Chronic stress, such as that which occurs during the disease, may convert these physiological stress granules into pathological stress granules, which can lead to the abnormal sequestration of SFPQ and nuclear tau in the cytoplasm resulting in an overall depletion from the nucleus. Of note, co-localization of SFPQ with oligomeric tau indicates a potential role of SFPQ in oligomerization and misfolding of the tau protein, which appears as a major hallmark of AD.

The significant reduction in SFPQ levels observed after human tau expression (tau-pathology model) *in vitro* and in the postmortem brains of rpAD subjects, suggests a causal role of tau in the downregulation of SFPQ. Quantitative proteomic analysis using Sequential Window Acquisition of all THeoretical fragment ion spectra-MS (SWATH-MS) in combination with functional characterisation illustrated two major themes (global translation reduction and DNA repair) that were altered as a consequence of the combinatorial effect of tau toxicity and SFPQ downregulation in this tau-pathology model.

Finally, the transgenic 3xTg-AD mice model uncovered specifically pre-symptomatic changes of target proteomic signatures. The levels of SFPQ and TIA-1 were already significantly elevated at an early pre-symptomatic phase of the disease in 3xTg-AD mice, suggesting that these proteins could be of potential significance as early therapeutic targets. This upregulation of the two SG components SFPQ and TIA-1 indicates active functions of the SG machinery at the early pre-symptomatic stage of the disease implicating pre-tangle stress, which coincides well with the observed acute phase oxidative stress-mediated upregulation of phospho-tau, TIA-1, and SFPQ in our cellular model of stress. Furthermore, the parallel reduction of SFPQ, which was found in the late symptomatic stage in 3xTg-AD mice and in the postmortem brains of patients with rapidly progressive forms of dementias (rpAD and prion disease), suggests that SFPQ may function as a common marker associated with rapid progression of these diseases.

On the basis of the findings from the current study, it can be concluded that the dislocation and dysregulation of SFPQ and nuclear tau, the subsequent DNA-related anomalies and aberrant dynamics of SGs in association with pathological tau represents a novel pathway which contributes to rapid progression in AD. Early pre-symptomatic changes in SFPQ indicate its relevance as an early therapeutic target. Reestablishing nuclear localization/expression of SFPQ might be a promising strategy to rescue neurodegeneration or to slow down the progression of the disease.

1 Introduction

1.1 RNA-binding proteins (RBPs)

The post-transcriptional regulatory mechanisms of neuronal gene expression are fast and effective processes that fine-tune the proteome of a cell in the brain to an ever-changing microenvironment (Glisovic *et al.*, 2008; Janga and Mittal, 2011; Richter and Klann, 2009; Wang and Szaro, 2016). These gene regulatory mechanisms are controlled by a group of special proteins known as RNA-binding proteins (RBPs) (Glisovic *et al.*, 2008; McMahon *et al.*, 2016). RNA-binding proteins are key regulators in RNA processing and translational control as they are complementary for RNAs, regulating all aspects of RNA metabolism; this includes alternative splicing, packaging, transport, stabilisation, translation, degradation, and the facilitation of RNA-interactions with other macromolecules (Fig. 1) (Anderson and Kedersha, 2009; Marchese *et al.*, 2016). More than 1500 RBPs have been described in the human cell (Castello *et al.*, 2012; Gerstberger *et al.*, 2014; Hentze *et al.*, 2018). RNA-binding proteins interact with *cis*-regulatory elements in the mRNA to form ribonucleoprotein (RNP) complexes, also known as granules (Wahl *et al.*, 2009), thereby controlling the function/expression of their target RNAs (Fig. 1) (Zhou *et al.*, 2014).

All RNP granules are composed of RBPs associated with mRNAs in their untranslated regions (5-´ or 3-´UTR) or in the coding regions (Anderson and Kedersha, 2009; Martin and Ephrussi, 2009). Transport granules are responsible for localization and storage of mRNAs for localized protein synthesis (Ramaswami *et al.*, 2013; Xing and Bassell, 2013). Stress granules (SGs) are reversible membrane-less aggregates in the cytoplasm which are formed in response to different environmental stresses (e.g. heat shock, oxidative stress, nutrient deprivation etc.); they are then resolved after the removal of stress (Anderson *et al.*, 2015; Guo and Shorter, 2015; Jain *et al.*, 2016). These granules have a pivotal role in stress response, sequestering mRNAs to sort them for their storage or decay (Jain *et. al.*, 2016). Degradation of mRNA occurs in processing bodies that are made up of proteins involved in degradation, surveillance of mRNA, repression of translation, and RNA-dependent silencing processes (Fig. 1) (Guo and Shorter, 2015; Jain and Parker, 2013; Marchese *et al.*, 2016).

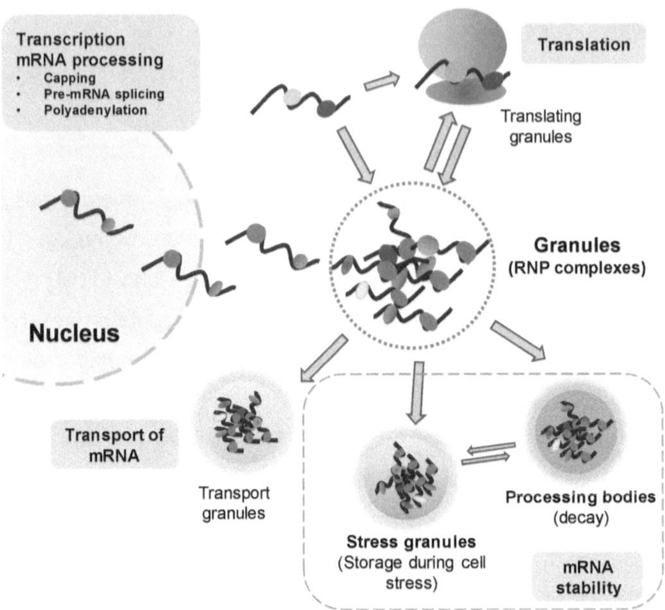

Figure 1: Functions of RNA-binding proteins through RNA-granule assembly. This figure highlights different steps of the RNA life cycle which are controlled by RBPs (grey rectangles). In the nucleus, RBPs are responsible for regulation of transcription, pre-mRNA processing, and export of newly synthesized mRNAs from the nucleus. Several types of granules present in the cytoplasm with their functions are described here. RNA granules are composed of an array of different proteins including RBPs, ribosomal subunits, helicases, translation factors and decay enzymes (Anderson and Kedersha, 2006). Disordered domains of RBPs contribute to dynamic remodeling of these granules by creating landing platform for RNA-protein and protein-protein interactions. Translating granules are responsible for translational control of the RNA. Stress granules safeguard mRNA and store them during stress response. Processing bodies are responsible for mRNA decay. Both stress granules and processing bodies are dynamic structures and share their components. Transport granules mediate the translocation of mRNAs through the long axons in neurons for *de novo* protein synthesis at the synapse (adapted and modified from Coppin *et al.*, 2018).

The interaction between RBPs and RNA is achieved by special domains in RBPs, known as RNA-binding domains (RBDs). RNA-binding proteins bind with RNA either by recognizing specific sequence elements in the target transcript or by recognizing secondary structural features in the RNA molecule (Castello *et al.*, 2016; Wright and Dyson, 2015). Therefore, binding specificity is dependent on both the secondary structural features of the RNA and the bound RBPs (Ding *et al.*, 2014; Gosai *et al.*, 2015; Li *et al.*, 2012). Typically, binding sites for RBPs are present in the untranslated regions of RNA but some sites can also be present in the coding regions as well

(Dassi *et al.*, 2012; Pickering and Willis, 2005). Sequence elements in the 5′-UTRs control the expression pattern of proteins, e.g. ribose-methylation of the cap, 5-terminal polypyrimidine sequences, and secondary structural elements. Sequences in the 3′-UTR region are responsible for regulation of stability, translational control, and localization (Lunde *et al.*, 2007; Wurth, 2012).

RNA recognition is complicated by the flexibility in the structure of the RNA. A single-stranded RBP might have strong sequence preferences, but the accessibility of an individual strand can depend on surrounding RNA structures (Duss *et al.*, 2014; Helder *et al.*, 2016). Furthermore, posttranslational modifications of RBPs are crucial, not only in diversifying their RNA-binding specificities but also in intracellular localization and metabolic functions (Glisovic *et al.*, 2008). Dysfunctional RBPs are emerging as key players in many neurological diseases (Maziuk *et al.*, 2018).

According to classical assumptions, RNA-protein interactions are dependent on well-defined, ordered globular domains. These well-structured RNA-binding domains are categorized into four main families: the zinc-finger domain (Brown, 2005), the K-homology domain (Valverde *et al.*, 2008), the RNA recognition motif (Clery *et al.*, 2008), and the double-stranded RNA-binding domain (Banerjee and Barraud, 2014). Until recently, RNA-protein interactions were assumed to be mediated mainly by these classical domains; however, new research, which characterizes other types of RNA-binding domains, has added more complexity to the intricate mesh of RNA-protein complexes (Balcerak *et al.*, 2019).

The binding of RBPs with RNA is not only specific but can also be non-specific, through auxiliary domains which greatly increase their functional diversity. Auxiliary domains consist of intrinsically disordered regions, which are composed of repetitive sequences of characteristic amino acids and a low percentage of hydrophobic amino acids. These domains enable the RBPs to form dynamic disordered structures ranging from collapsed globules to extended coils (Dyson and Wright, 2005; Varadi *et al.*, 2015; Wright and Dyson, 2015). Intrinsically unstructured regions in the RNA-binding proteins are important in two ways. Firstly, these segments establish extended yet conserved electrostatic boundaries with RNAs via induced fit. Secondly, flexibility in their conformation enables them to bind different RNA targets, providing multi-functionality while also ensuring specificity (Varadi *et al.*, 2015; Calabretta and

Richard, 2015). These regions are responsible for reversible phase transition, leading to formation of liquid droplets, hydrogels, and aggregates or fibrils (Brangwynne, 2013; Hyman *et al.,* 2014). RNA-protein interactions mediated by auxiliary domains affect many aspects of RNA-processing; their disruption, therefore, can potentially cause protein disorders (Varadi *et al.,* 2015; Calabretta and Richard, 2015).

An important subset of low-complexity domains are prion-like domains (PLDs), mainly comprised of uncharged polar residues and glycines, showing similarities with the prion protein of yeast (Couthouis *et al.,* 2011; Lancaster *et al.,* 2014). These PLDs enable several proteins of yeast, e.g. Sup35 and Rnq1, to form infectious structures, termed prions (Alberti *et al.,* 2009; King *et al.,* 2012; Toombs *et al.,* 2010; Wickner *et al.,* 2015). Deletion of these prion domains precludes access to the prion state (Masison *et al.,* 1997), and the addition of this region to otherwise innocuous proteins is sufficient to induce prion-like behaviour (Li and Lindquist, 2000; Tyedmers *et al.,* 2010).

The development of bioinformatics algorithms has led to the identification of *bona-fide* prion domains (Alberti *et al.,* 2009; Couthouis *et al.,* 2011; King *et al.,* 2012; Toombs *et al.,* 2010). These algorithms scan amino acid compositions to screen the human genome for proteins with PLDs. One of the updated PLD detection algorithm is PLAAC (Prion-Like Amino Acid Composition) (Lancaster *et al.,* 2014), which has been used for PLD prediction for several organisms.

There are about 240 human proteins with PLDs that have been identified by PLAAC. Of these, 70 are RBPs, suggesting a beneficial and essential role of these domains. Prion-like domains are essential for RBP functions and enable them to undergo liquid-liquid phase separation (LLPS). This phase separation is the basis for the formation of higher-order structures, including oligomers and several membrane-less granules (Fig. 2) (Toretsky and Wright 2014; Verdile *et al.,* 2019). However, this LLPS property renders prion-like-domain-containing proteins prone to misfold and aggregate via aberrant phase transitions (Fig.2) (Harrison and Shorter, 2017; Verdile *et al.,* 2019). Prion-like domains form mesh-like networks *in vitro*, manifesting as hydrogels (Kato *et al.,* 2012). These hydrogels are different from amyloid material and signify a functional amyloid (Hennig *et al.,* 2015). Due to this special property of functional aggregation, PLD-containing proteins have gained attention in recent years in protein aggregation disorders, e.g. amyotrophic lateral sclerosis (ALS), Alzheimer's, and prion diseases (Harrison and Shorter, 2017; March *et al.*, 2016; Wolozin, 2012).

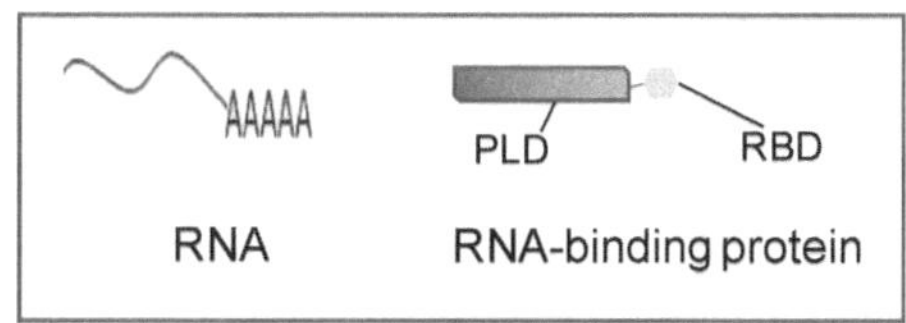

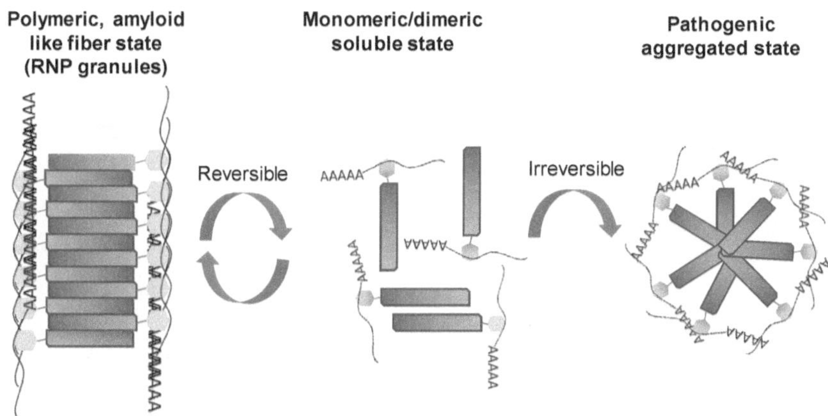

Figure 2: **Prion-like granule assembly by RNA-binding proteins with prion-like domains**. The prion-like domains enable RBPs to exist in one of three states: a soluble state (monomeric), an amyloid-like fiber state (polymeric) or a pathogenic (aggregate) state. This polymeric amyloid state is the basis of formation of different types of granules, e.g. stress granules, transport granules, and processing bodies. The conversion between the first two states is reversible, which means RBPs can both enter and exit a prion-like state. Transition to 3rd state is pathogenic and is irreversible (Kato *et al.*, 2012) (adapted from Gao and Arkov, 2013).

Regulation of RNA-biology is highly complex, due to the heightened demands of RBP functions in the neurons. RNA-binding proteins are important for neurons in two distinct ways. Firstly, alternative splicing is particularly active in the neurons, as compared with other tissues (Li *et al.*, 2007; Yeo *et al.*, 2004). This posttranscription-al regulatory mechanism (alternative splicing) is dependent on RBPs. Secondly, it is RBPs that ensure safe transportation of mRNAs from the nucleus to the cytoplasm, dendrites and long axons, thus preventing their premature degradation and transla-tion during their journey (Fig. 3) (Anji and Kumari, 2016; Holt and Bullock, 2009; Zhang and Poo, 2002; Zhou *et al.*, 2018). In addition, RNA-binding proteins are effi-cient and fast regulatory hubs, helping the neuron to manage the strains of an everchanging microenvironment, including synaptic depolarization responses, depression, oxidative stress, misfolded proteins, reduced nutrient availability, and apoptosis (Sephton and Yu, 2015; Zhou *et al.*, 2018).

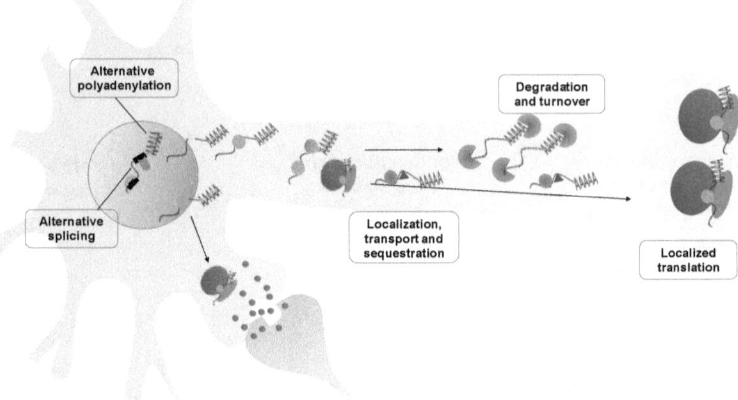

Figure 3: RNA-binding proteins in the neuron. RNA-binding proteins are very important in the neurons in order to fulfill heightened demand of dynamic RNA-RBP processes in the neurons. Firstly, alternative splicing is particularly active in the neurons compared with other tissues. Secondly, neurons need RBPs to transport mRNAs through long axons to distal parts of the neurons, e.g. the synapse for *de novo* protein synthesis. As a result, dysfunction of the RNA-binding proteins leads to defects in post-transcriptional gene regulatory mechanisms, resulting in neurodegenerative disorders (adapted and modified from Zhou *et al.*, 2018).

In several neurodegenerative diseases, alterations in the dosage and dynamics of RBPs, including expressional changes, mutations, aberrant interactions or altered RNA-processing, are emerging as a major pathological feature (Castello *et al.*, 2013; Liu *et al.*, 2017; Maziuk *et al.*, 2018; Nussbacher *et al.*, 2015). Dysregulation of many RBPs, including FUS, TDP-43, hnRNPA1 and ATXN2, have been demonstrated in amyotrophic lateral sclerosis, frontotemporal lobar degeneration and/or spinocerebellar ataxia (Table 1) (Maziuk *et al.*, 2017). Other RBPs have also been shown to co-aggregate with aggregated-prone protein inclusions in AD, Huntington's disease (HD) and Creutzfeldt-Jakob disease (CJD) (Table 1) (Goggin *et. al.*, 2008; Maziuk *et al.*, 2017; Nussbacher *et al.*, 2015; Zhou *et al.*, 2014).

Table 1: Different RBPs associated with neurological diseases.

RNA-binding proteins	Associated diseases
TAR DNA-binding protein 43 (TDP-43)	ALS, FTLD, AD, HD
T-cell intracellular antigen 1 (TIA-1)	ALS, FTLD, AD
Ras GTPase-activating protein-binding protein 1 (G3BP1)	ALS, FTLD, AD
Tristetraprolin (TTP)	ALS, FTLD, AD
Fused in Sarcoma (FUS)	ALS, FTLD
Ewing Sarcoma protein (EWS)	ALS, FTLD
TATA-Box Binding Protein Associated Factor 15 (TAF15)	ALS, FTLD
Heterogenous Ribonucleoprotein Particle A1/A2 (hnRPA1/A2)	ALS, FTLD
Angiogenin (ANG)	ALS, PD
Survival of motor neuron (SMN1)	ALS, SMA
Matrin-3 (MATR3)	ALS
Ataxin-2 (ATXN2)	ALS
Optineuin (OPTN)	ALS
Fragile X mental retardation protein (FMRP)	FXS

ALS: Amyotrophic lateral sclerosis, FTLD: Frontotemporal lobar degeneration, AD: Alzheimer's disease, HD: Huntington's disease, PD: Parkinson's disease, FXS: Fragile X syndrome, SMA: Spinal muscular atrophy (Maziuk *et al.*, 2017).

One by one, connections are being discovered between RNA-binding proteins with prion-like domains and neurodegenerative disorders (Li *et al.*, 2013; March *et al.*, 2016). Recently, this paradigm of RNA-binding proteins has been extended to AD, where pathological aggregates of TIA-1 (TIA-1: cytotoxic granule-associated RBP), an RNA-binding protein with prion-like domain, have been linked to tau neurofibrillary tangles (NFTs) in the brain (Vanderweyde *et al.*, 2012, Vanderweyde *et al.*,2016). Furthermore, co-aggregation of components of the spliceosomal complex with the tau protein has been reported in both sporadic and familial AD cases, but not in other tauopathies (Bai *et al.*, 2013; Bishof *et al.*, 2018; Diner *et al.*, 2014; Sengupta *et al.*, 2018).

1.2 Alzheimer's disease

Alzheimer's disease is the most prevalent cause of dementia with progressive neurodegeneration, affecting 40–50 million people around the globe (GBD Dementia Collaborators, 2019; Prince *et al.,* 2013). This number is predicted to increase threefold by 2050 (Prince *et al.*, 2013; Prince *et al.*, 2015). The prevalence of AD is quite high in North America and Europe, compared with less developed countries, though a sharp increase has been observed in India, China and Latin America in recent years (Ferri *et al.*, 2005; Kalaria *et al.*, 2008). In America alone, the financial burden of maintaining AD patient's health care amounts to ~$203 billion annually. Due to a total lack of therapeutic interventions for the treatment and prevention of AD, the costs are expected to reach $1.1 trillion annually by the year 2050 (Alzheimer's Association, 2019).

Alzheimer's disease has two sub-classifications: early-onset (EOAD) or familial AD (FAD), or late-onset (LOAD) or sporadic AD (SAD). For FAD, symptoms usually appear earlier than SAD, typically ranging between 30-50 years of the age (Bertram *et al.*, 2010; Goate and Hardy, 2012; Sanabria-Castro *et al.*, 2017). Autosomal dominant mutations in amyloid-precursor protein (APP), presenilin-1, and -2 genes have been shown to cause familial AD (Blennow *et al.*, 2006). Known genetic causes of AD only account for a small percentage of cases (less than 1%). Alzheimer's disease cases, which are not associated with any genetic mutations, are known as sporadic AD (Mendez, 2017).

Clinically, AD is characterized by continuous memory deficits and dysfunction of other cognitive abilities. At initial phases, the major symptoms are centered on episodic memory. With progression of the disease, topographical difficulties emerge, alongside problems with multi-tasking and loss of confidence. By the time a patient is diagnosed with AD dementia, symptoms have typically become more sever, interfering with activities of daily life (Scheltens *et al.*, 2016). At later stages of the disease, other deficits may also emerge, including impaired mobility, behavioural abnormalities, hallucinations and delusions. Severe stages of the disease are accompanied by a complete loss of various cognitive functions, impaired motor functions (e.g. chewing and swallowing) and linguistic problems. Most of the patients are bedridden at this stage, and die of inanition or secondary illnesses, e.g. infections and ulcers (Förstl and Kurz, 1999; Tarawneh and Holtzman, 2012).

Neuropathologically, AD is characterized by two cardinal hallmarks: intracellular tangles of misfolded tau protein in conjunction with extracellular plaques of aggregated amyloid-β (Aβ) peptide (Cushman *et al.*, 2010; Perl, 2010), together with neuronal and synapse loss (Nelson *et al.*, 2009; Selkoe and Hardy, 2016; Perl, 2010). The Aβ plaques originate from the aggregation of Aβ peptides (40–42 amino acid long), produced by sequential cleavage of APP by the β- and γ-secretases. An imbalance between the production and clear mechanisms for Aβ peptides leads to precipitation of Aβ pathology (Selkoe and Hardy, 2016).

The amyloid hypothesis posits Aβ pathology as the primary pathological feature of the disease (Hardy and Higgins, 1992), triggering a cascade of further pathological events, including the formation of neurofibrillary tangles of hyperphosphorylated tau, neuroinflammation, oxidative stress, and neuronal loss (Hardy and Higgins, 1992; Reitz, 2012; Tanzi and Bertram, 2005). Unfortunately, therapeutic interventions targeting Aβ have failed to improve cognitive functions in AD (Doody *et al.*, 2014; Lovestone *et al.*, 2015). One possible reason for the failure of these therapeutic strategies is an incomplete understanding of the mechanisms leading to neurodegeneration in AD. Mounting evidence has shifted the focus towards tau as a more promising therapeutic target for AD (Cao *et al.*, 2018). The discovery of pathogenic mutations in the tau gene in familial cases of frontotemporal dementia (FTLD-17: FTLD with parkinsonism linked to chromosome 17) has demonstrated a clear link between the dysfunction of tau and neurodegeneration (Goedert *et al.*, 2000).

Tau, suggested by some to be the "holy grail of dementia," is a protein initially described as a dull executor of pathological effects associated with amyloid β. In AD, tau is hyperphosphorylated, misfolded, oligomerized, aggregated, and mislocalized (Grundke-Iqbal, 1986; Ren and Sahara, 2013; Vanderweyde *et al.*, 2016). Since the initial discovery of tau in 1975, the field has focused on its role in microtubule stabilization by binding with polymerized tubulin in the axons (Weingarten *et al.*, 1975). Over the last few decades, several studies have reported multiple functions and localizations of tau protein. Specifically, its localization in the nucleus (both phosphorylated and non-phosphorylated) (Bukar *et al.*, 2016) and cytoplasm under conditions of oxidative stress has been demonstrated (Vanderweyde *et al.*, 2016). Unfortunately, the significance of multiple localization types is not yet clear.

Recent evidence suggests a novel pathological feature of tau in relation to cytoplasmic stress granules, through which tau disrupts cellular homeostasis. RNA-binding proteins, like TIA-1, co-localize with hyperphosphorylated tau and aggravate tau pathology (Fig. 4) (Vanderweyde *et al.*, 2012; Vanderweyde *et al.*, 2016). This interaction between tau and TIA-1 in stress granules has been shown to enhance tau-mediated neurodegeneration in primary hippocampal cultures, which can be rescued by reduction of TIA-1 protein (Fig. 4) (Vanderweyde *et al.*, 2016). Likewise, extracellular tau, after internalization and hyperphosphorylation, has been shown to alter SG dynamics, supporting the notion that secreted tau has a role in the formation of pathological SGs (Brunello *et al.*, 2016). All these evidences link tau pathology to dysfunctional RBPs and pathological stress granules, highlighting an important role of RBPs in AD.

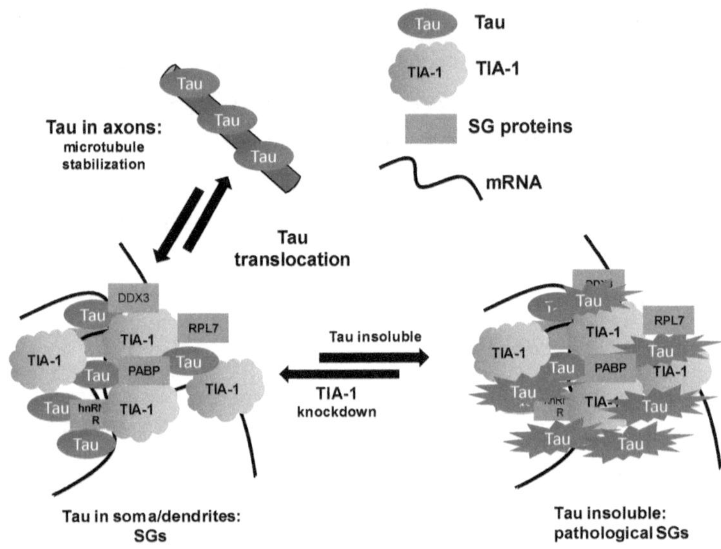

Figure 4: The interplay between tau and RNA-binding proteins in the SGs: Tau is normally present in the axons, but stress induces translocation to somatodendritic compartments. RNA-binding proteins, e.g. TIA-1 with both nuclear and cytoplasmic functions, keep on translocating between cytoplasmic and nuclear regions. Although TIA-1 is predominantly a nuclear protein, stress leads to rapid shuttling into the cytoplasm, where its interaction with tau promotes SG formation. This association also enhances tau misfolding and aggregation and increases the size of SGs by influencing the RNA-binding protein composition of SGs. Furthermore, this contact increases the tendency of tau to form sarkosyl-insoluble aggregates and stabilize SGs. These deleterious effects can be rescued by TIA-1 reduction in cultured neurons (adapted and modified from Vanderweyde *et. al.*, 2016).

The current neuropathological assessment of AD is based on updated criteria released by the National institute of Aging (NIA). Classification of AD neuropathologic changes is achieved according to three different staging themes: Thal stages for distribution of amyloid-β deposits (Thal *et al.*, 2002), neurofibrillary tangle pathology with Braak stages (Braak and Braak, 1991; Braak *et al.*, 2006), and the occurrence and severity of neuritic plaques according to the Consortium to Establish a Registry for Alzheimer's Disease (CERAD) (Fillenbaum *et al.*, 2008; Mirra *et al.*, 1991). Combining these three standards (Amyloid: A, Braak: B, CERAD: C) to the ABC method provides an estimate of a no, low, medium or high pathology (Montine *et al.*, 2012).

The genetic and non-genetic risk factors associated with AD are significant, because they give clues into the predispositions of the disease process prior to onset. Furthermore, they provide basis for classification of individuals with increased risk for the disease. Diverse non-genetic risk factors have been linked with AD including cardiovascular diseases, hypertension, type 2 diabetes, obesity, traumatic injury to the head, life style (poor diet, physical inactivity, smoking etc.), and depression (Crous-Bou *et al.*, 2017; Edwards *et al.*, 2019; Reitz and Mayeux, 2014).

Genetic risk factors also affect sporadic AD. Several genes have been identified as a risk factor for sporadic AD including *TREM2, PLD3, ADAM10, CD2AP, DSG2,* and *APOE* (Karch and Goate, 2015). Among these genes, *APOE* polymorphism has been one of the most widely studied risk factor. The *APOE* gene containing three variants (ε2, ε3 and ε4) represents the greatest risk for developing sporadic AD. Here, ε4 is the main risk factor, as ε4 heterozygotes have three-fold higher risk, which rises to twelve-fold in ε4 homozygotes (Karch and Goate, 2015; Mahley, 2016). Given the low percentage of genetic causes, the identification of potential environmental risk factors for sporadic AD is crucial.

Although age is considered a key risk factor for AD, several studies have also implicated chronic stress as a crucial environmental risk factor (Hoeijmakers *et al.*, 2017; Huang *et al.*, 2016; Islam *et al.*, 2019; Justice, 2018; Lesuis *et al.*, 2016; Ownby *et al.*, 2006). It has been proposed that oxidative stress plays an important role in the initiation and progression of AD (Wang *et al.*, 2014). Oxidative stress is caused by an imbalance in the biochemical systems involved in the production and removal of reactive oxygen species (ROS). Reactive oxygen species are reactive molecules originated from oxygen (Andreyev *et al.*, 2005), which are highly reactive due to presence of their unpaired electrons (Patten *et al.*, 2010).

Several studies clearly indicate that an increased level of reactive oxygen species leads to oxidative stress and the manifestation of neurodegenerative disorders, including AD, ALS, and PD (Chen and Liu, 2017; Magalingam *et al.*, 2018; Niedzielska *et al.*, 2016; Patten *et al.*, 2010). Oxidative stress can affect cell biology in many different ways, including damage to cell membranes and other functional units such as proteins, lipids, and DNA. In addition, oxidative stress may result in aberrant dynamics of SGs (Chen and Liu, 2017). The brain is particularly vulnerable to these insults because of its heightened demand for oxygen and its low antioxidant capacity (Chen *et al.*, 2012; Cobley *et al.*, 2018). Pathological and persistent SGs due to chronic stress and mutations in RBPs have been implicated in several neurodegenerative diseases (Liu-Yesucevitz *et al.*, 2014; Vanderweyde *et al.*, 2016). However, whether these RBP-pathological features are associated with rapid progression of the disease remains enigmatic.

1.3 Atypical subtype of Alzheimer's disease

Typically, sporadic AD is characterized by a slow progression in cognitive decline, with conspicuous memory loss. Classically, disease duration spans ~8 years after the onset of clinical symptoms (Scheltens *et al.*, 2016). However, emerging evidence shows heterogeneity in both clinical phenotypes and progression rates (Abu-Rumeileh *et al.*, 2018; Ba *et al.*, 2017; Cohen *et al.*, 2015; Schmidt *et al.*, 2011). These classical AD cases are abbreviated as spAD in the current study.

Recently, a rapidly progressive variant of AD (rpAD) has been described with a steep decline in the Mini-Mental State Examination (MMSE: a psychometric test) score (e.g. $\geq$ 6 points/year), and/or a reduced survival duration (~4 years in comparison to 8 years for spAD cases) (Llorens *et al.*, 2016; Nelson *et al.*, 2009; Schmidt *et al.*, 2010; Soto *et al.*, 2008; Tosto *et al.*, 2015). The prevalence and clinical definition of rpAD varies greatly across different studies in the literature. Preliminary evidence supports the notion that rpAD is associated with a specific molecular and pathogenic cascade (Ba *et al.*, 2017; Cohen *et al.*, 2015; Drummond *et al.*, 2017; Schmidt *et al.*, 2011). However, no significant differences have been detected in the core neuropathological features between spAD and rpAD (Cohen *et al.*, 2015; Schmidt *et al.*, 2010), suggesting a great demand for a better understanding of molecular signatures responsible for this heterogeneity.

Rapidly progressive AD also exhibits significant clinical overlap with Creutzfeldt-Jakob disease, another rapidly progressive dementia, which makes early differential diagnosis a challenge. The rpAD often mimics the disease duration of CJD. Previous reports have also demonstrated similarity in biomarker profile and clinical features (rapid course, development of early focal neurological signs, levels of CSF markers

14-3-3 and total tau) between rpAD and CJD (Abu-Rumeileh *et al.*, 2017; Schmidt *et al.*, 2011; Stoeck *et al.*, 2014).

In fact, in prion disease referral centres worldwide, rpAD is a common non-prion diagnosis at neuropathological investigation, accounting for ~14–50% of all non-CJD cases (Chitravas *et al.*, 2011; Jansen *et al.*, 2012; Lattanzio *et al.*, 2017; Stoeck *et al.*, 2012). The lack of understanding, how molecular mechanisms and risk factors lead to rapid progression seen in the various rapidly progressive forms of dementia (rpAD and sCJD), has hindered the development of therapeutic interventions, specifically at the early stages.

The most common human prion disease, CJD is a rapidly progressive, rare, transmissible and fatal disease, with patients exhibiting dementia among other major clinical symptoms (Collinge, 2001; Johnson and Gibbs, Jr., 1998; Prusiner, 1982; Zerr and Parchi, 2018). There are four major types of CJD: sporadic, familial, iatrogenic and the variant form. Sporadic CJD is the most prevalent form (85%) of all human prion diseases (Parchi *et al.*, 1999; Tschampa *et al.*, 2007; Zerr and Parchi, 2018). Within sCJD, there are six distinct subtypes as determined by a combination of polymorphism at codon 129 of prion protein gene *(PRNP)* and two types of strains (type 1 and 2). The six subtypes have been classified as sCJD-MM1, -MV1, -VV1, -MM2, -MV2 and -VV2 corresponding to different clinical phenotypes (Bishop *et al.*, 2010; Parchi *et al.*, 1999; Parchi *et al.*, 2009). Among these subtypes, sCJD-MM1 and sCJD-VV2 subtypes are the most prevalent ones (Parchi *et al.*, 1999; Meissner *et al.*, 2009), therefore they were investigated in the present study along with AD-subtypes to uncover common molecular factors underlying variant progression rates.

During the last decades, there has been significant progress in understanding the risk factors and molecular basis underpinning AD. It has become clear that AD is a multifaceted disease; multiple theories have been proposed, with the amyloid-cascade hypothesis being the most studied one. The recent failure of many Aβ-oriented clinical trials has disappointed the field and led to a focus on other molecular mechanisms associated with this complex neurodegenerative disorder. Furthermore, recent discovery of various progression phenotypes of AD demands a great need for understanding molecular factors, leading to heterogenous progression phenotypes in AD. Emerging evidence indicates an increasingly compelling role of dysfunctional neuronal RBPs and stress granules (Ramaswami *et al.*, 2013, Wolozin,

2012) in neurodegenerative diseases. To this end, this study aims to identify and characterize RNA-binding proteome (RBPome) alterations in subtypes of Alzheimer's and prion diseases, to open new avenues for early diagnosis and disease-modifying therapies.

1.4 Objectives of the study

Based on previous observations, we hypothesize that global deregulation of RNA-RBP processes contributes to the pathophysiology of Alzheimer's and other neurodegenerative diseases, particularly prion diseases. RNA-binding proteins may represent a previously "hidden" component of pathophysiology of Alzheimer's and prion diseases. Understanding the global derangement of RBPs during the course of the disease will potentially unveil new targets for the design of therapeutics.

The present study was conducted to uncover pathological mechanisms linked to dysfunctional RBPs, leading to heterogeneous progression rates and phenotypes of AD. The study principally focusses on defining differential RBPome signatures in spAD, rpAD, two prion disease subtypes (sCJD-MM1 and sCJD-VV2) and age-matched controls. In the next part of the study, target proteomic candidates will be characterized in the human brain as well as cellular and animal models to explore the mechanistic role of these signatures in the pathogenesis and progression of the disease.

The objectives of the study were:

1. to identify and characterize RBPome signatures from diseased and healthy subjects, to have an overview of multiple deregulated pathways,
2. to perform a differential expressional analysis of target RBPs in the human brain and mouse brain model and at a cellular model,
3. to identify and characterize pathological mechanisms associated with target RBPs in the progression rate variations, and
4. to translate RBP signatures in cellular and animal models, to find a mechanistic link with the pathological features of the disease.

2 Materials and methods

2.1 *Materials*

2.1.1 Antibodies

All the antibodies used for immunoblotting (IB) and immunofluorescence (IF) are listed in Tables 2 and 3.

Table 2: List of primary antibodies used in the study

Primary Antibody	Origin	Dilution (IB)	Dilution (IF)	Cat. No./ Company
Tau-5	Mouse	1: 500	1: 100	ab80579/Abcam
Tau (E178)	Rabbit	1: 1000	1: 100	ab32057/Abcam
Anti-tau (T22), oligomeric	Rabbit	1: 1000	1: 250	ABN454/Sigma-Aldrich
Phospho-tau (S199)	Rabbit	1: 1000	1: 100	ab81268/Abcam
TIA-1	Rabbit	1: 500	1: 100	ab140595/Abcam
TIA-1	Mouse	1: 500	-	ab40693/Abcam
TIA-1(G-3)	Rabbit	1: 200	1: 100	sc-166247/Santa Cruz
TIA-1 (G-3) AlexaFluor 488	Mouse	-	1: 50	sc-166247/Santa Cruz
SFPQ	Rabbit	1: 500	1: 100	ab38148/Abcam
VCP	Rabbit	1: 3000	1: 200	ab109240/Abcam
GAPDH	Mouse	1: 3000	-	G8795/Sigma-Aldrich
β-Actin	Mouse	1: 1000	-	ab8227/Abcam
BRD4	Rabbit	1: 1000	-	ab128874/Abcam

Table 3: List of secondary antibodies used in the study

Secondary antibody	Origin	Dilution (IB)	Dilution (IF)	Cat. No. /Company
α-Mouse-HRP (IgG)	Goat	1: 10000	-	115-035-062/Jackson IR Lab
α-Mouse-HRP (IgM)	Goat	1: 10000	-	AP128P/Merck Millipore
α-Rabbit-HRP (IgG)	Goat	1: 10000/1: 5000	-	11-035-144/Jackson IR Lab
α-Mouse-A488	Goat	-	1: 200	A-11001 /Invitrogen
α-Rabbit-A488	Goat	-	1: 200	A-11008/Invitrogen
α-Mouse-A555	Goat	-	1: 200	A-21424/Invitrogen
α-Rabbit-A546	Goat	-	1: 200	A-11010/Invitrogen

2.1.2 Antibiotics, enzymes and standards

Table 4: All antibiotics, enzymes and standards used

	Cat. No. / Company
Antibiotics	
Ampicillin	171254, Calbiochem
Penicillin-Streptomycin (PS)	15140122, Thermo Fisher Scientific, Dreieich, Germany
Enzymes	
Taq DNA Polymerase, 5 U/µL	11146173001, Sigma-Aldrich, Deisenhofen, Germany
Standards (Protein and DNA)	
Bovine serum albumin (BSA)	P0914, Sigma-Aldrich
DNA ladder	SM1333, Thermo Fischer Scientific
Precision Plus Protein Standard	161-0374, Bio-Rad, Munich, Germany

2.1.3 Bacterial strain and culture media

Table 5: List of bacterial strain and culture media

Bacterial Stain and media	Catalog No./Company
E. coli strain DH5α	Addgene
LB medium	A0954/PanReacAppliChem ITW reagents
LB agar	A0949/ PanReacAppliChem ITW reagents

2.1.4 Cell culture reagents

Table 6: Reagents used in cell culture

Reagent	Catalog No./Company
DMEM, high glucose, HEPES, no phenol red	21063/Thermo Fisher Scientific
Fetal bovine serum (FBS)	F7524/Sigma-Aldrich
Lipofectamine 2000	11668027/Thermo Fisher Scientific
Opti-MEM, reduced Serum Medium, no phenol red	11058021/Thermo Fisher Scientific
Phosphate-buffered saline (PBS)	L1825/Merck
Trypsin/EDTA solution	T4174/Sigma-Aldrich
GlutaMAX supplement	Gibco 35050038/Thermo Fischer Scientific

2.1.5 Chemicals

All chemicals used in the present study were obtained from Sigma-Aldrich (Deisenhofen, Germany), Merck (Darmstadt, Germany), Roth (Karlsruhe, Germany), Bio-
Rad (Munich, Germany), Amersham (Freiburg, Germany), Fluka (Deisenhofen,

Germany), Thermo Fisher Scientific (Darmstadt, Germany), or unless otherwise stated.

2.1.6 Instruments and other materials

Table 7: Instruments and appliances used in the study

Appliances	Model/Description	Manufacturer
Centrifuges	5415C	Eppendorf, Hamburg, Germany
	Optima TL 100	Beckman, Krefeld, Germany
C1000 Touch Thermal Cycler		Bio-Rad, USA
ChemiDoc XRS+ system	170-8265	Bio-Rad
Electrophoresis apparatus	Mini-ProteanSarstedt III	Bio-Rad
Filtopur V50 0.2 (Vacuum filter)	83.1823.001	SARSTEDT, Nümbrecht, Germany
Heated magnetic stirrer	iKAMAG RCT	IKA-Labortechnik, Staufen, Germany
Ice machine	-	Ziegra, Isernhagen, Germany
Incubator	IFE 400	Memmert, Schwabach, Germany
Light Cycler 480 Multiwell Plate 96, white	-	4729692001, Roche Life sciences, Germany
Microscope	Leica TCS SPE	Leica Microsystems, Wetzlar, Germany
Microscope	Zeiss LSM 510 Meta	Carl Zeiss
Microscope	Zeiss 667183 Axiovert 25C	Carl Zeiss
Microwave oven	ER-6320 PW	Brother International, Bad Vilbel, Germany
Microplate reader	Perkin Elmer Wallac 1420 Victor	GMI, USA
Power supply	Power Pac 300	Bio-Rad
Safe-Lock tubes	0.2, 0.5, 1.5 and 2ml	Eppendorf
Semi-Dry transfer Cell	Transblot Turbo transfer system	Bio-Rad
Serological pipettes plastic tubes	2, 5, 10, 25ml 15 and 50ml	Sarstedt
pH meter	pH 526	WTW, Weilheim, Germany
pH strips (6.5-10)	1.09543.0001	Merck Millipore, Germany
Shakers	CERTOMAT R	Sartorius, Göttingen, Germany
Spectrophotometers	EL808	Bioteck instruments, Winooski-vermont, Germany
Syringes BD Discardit	2, 5, 20ml	Becton Dickinson, NJ, USA
TC-plate 6 well, Cell+F	83.3920.300	SARSTEDT
TC Flask T75, Cell+vented Cap	83.3911.302	SARSTEDT
Thermomixer	5436	Eppendorf
TissueLyser LT	85600	Qiagen, Hilden, Germany
UV-transilluminator	200x 200mm	Bachofer, Reutlingen, Germany
Vacuum drier	UNIVAPO 150H	UNIEQUIP, Martinsried, Germany

Water bath	1003	GFL, Burgwedel, Germany

2.1.7 Kits

Table 8: Kits used in the present study

Name	Cat No. / Company
Purelink Genomic DNA isolation Kit	K182001, Invitrogen
RNeasy Plus Universal Mini Kit	73404, Qiagen
HI Speed Plasmid Midi Kit	12643, QIAGEN
Pierce Magnetic RNA-Protein Pull-Down Kit	20164, Thermo Fisher Scientific
MTS Assay Kit (Cell Proliferation) (Colorimetric)	ab197010, Abcam
Chemiluminescent Nucleic Acid Detection Module Kit	Thermo Fisher Scientific
High-Capacity cDNA Reverse Transcription Kit	4368814, Thermo Fisher Scientific

2.1.8 Mammalian cell lines and culture media

2.1.8.1 HeLa cells: HeLa cells were kindly provided by Dr. Aman-Deep Singh Arora, European Neuroscience Institute Göttingen, Georg-August University Göttingen, Germany. The cells were cultured in DMEM, supplemented with 10% FBS and 1% PS at 37°C with 5% CO_2 and 95% humidity.

2.1.8.2 SH-SY5Y cells: SH-SY5Y cells were obtained from Prof. Walter Schulz-Schaeffer, Department of Neuropathology, University Medical Center (UMG), Göttingen, Germany. The cells were cultured in DMEM, supplemented with 10% FBS, 1% GlutaMax supplement and 1% PS at 37°C with 5% CO_2 and 95% humidity.

2.1.9 Plasmids

Plasmids for human wild-type tau (pRK5-EGFP-tau, cat. #46904) and mutated tau (pRK5-EGFP-tau P301L, cat. #46908) were purchased from Addgene (originally prepared by Karen Ash lab) (Hoover *et al.*, 2010).

2.1.10 Primer pairs

All primers were purchased from Eurofins Genomics. List and sequences of primer pairs are provided in annexure data Table 14.

2.1.11 Software and online tools

Table 9: List of software and web-based tools

Name	Description/use	References
GraphPad Prism (6.01)	Statistical analysis	GraphPad Software, Inc. California, USA
catGRANULES	Liquid-Liquid phase separation property (LLPS) estimation	http: //s.tartaglialab.com
Functional enrichment analysis tool (FunRich)	Functional enrichment analysis	http: //www.funrich.org/
FIJI 1.52p	Statistical analysis	National institutes of Health, USA
Image J 1.51j8	Immunofluorescence analysis	National institutes of Health, USA
Image Lab (3.0.1)	Densitometric analysis	Kapelan, GmbH/Halle, Germany
Inkscape (0.92)	Professional quality vector graphics software	https: //www.inkscape.org
IPA	Pathway mapping	Qiagen, USA
LAS X	Imaging software	Leica Microsystems/Wetzlar, Germany
Perseus software (1.5.0.31)	Proteomics data analysis	MPI of Biochemistry, Martinsried, Germany
PLAAC software	Prion-like domain scanning	http: //plaac.wi.mit.edu
R version 3.4.3	Statistical analysis	Proteome Software, Inc
Cytoscape 3.6.1	Protein network visualization	Cytoscape.Js
Scaffold 4.8.4	MS/MS data analysis	Proteome Software, Inc
WEB-based GEne SeT AnaLysis Toolkit (WebGestalt)	Functional enrichment analysis	http: //www.webgestalt.org/
Zeiss LSM 4.2.0.121	Immunofluorescence imaging	Microimaging GmbH, Göttingen, Germany

2.1.12 Solutions and buffers

Note: ddH_2o water was used to prepare solutions and buffers

Blocking solution for immunoblotting: 5% non-fat dry milk in PBS-T/TBS-T

Cell-lysis buffer: 50 mM Tris-HCl, pH 8, 1% Triton X-100, 0.5% CHAPS, 1 mM DTT

Citrate buffer: 10 mM sodium citrate (pH 6.0)

Coomassie stain: 0.1% Coomassie blue R-250, 10% acetic acid, 50% methanol, 40% ddH_2O

PBS-T buffer: PBS and Tween-20 (0.05% Tween)

Resolving gel buffer: 1.5 M Tris, 0.4% SDS, pH 8.8

SDS-running buffer: 25 mM Tris, 192 mM glycine, 0.1% SDS, pH 8.3

Stacking gel buffer: 0.5 M Tris, 0.4% SDS, pH 6.8

TBS-T buffer: 50 mM Tris, 150 mM NaCL, 0.05% Tween-20, pH 7.6

Tissue lysis buffer: 7 M urea, 2 M thiourea, 4% CHAPS (freshly added 2% ampho-lytes, 1% DTT)

Transblot-buffer: 192 mM glycine, 10% methanol, 25 mM Tris-HCl, pH 8.3

2.2 Methods

2.2.1 Patient cohorts and sample processing

Patient cohort processing, neuropathological examination and brain tissue collection for this study were all conducted as previously described (Krbot and Glatzel, 2018; Zafar *et al.*, 2018; Zafar *et. al.*, 2017). Briefly, postmortem brain material was obtained from patients after the approval of the local ethics committee at the University Medical Center, Göttingen, Germany. Frontal cortex samples from spAD, rpAD and age-matched controls were provided by the brain bank of the Institute of Neuropathology (HUB-ICO-IDIBELL Biobank) and Biobank of Hospital Clinic-IDIBAPS Spain, following the legislation (Ley de la Investigación Biomédica 2013 and Real Decreto-Biobancos, 2014). Frontal cortex samples were obtained from patients with sCJD subtypes (MM1 and VV2), from the Department of Neurology at the University Medical Centre, Göttingen.

Cases strictly fulfilling the following inclusion criteria were included in rpAD cohort:

1. initial classification as prion disease based on rapidly progressive neurological and clinical parameters;

2. presence of AD pathological features, i.e., higher Braak stages and CERAD stages;

3. exclusion of other rapidly progressive dementias (e.g. prion disease) and potential causes of rapid progression, e.g. vascular pathology, extensive Lewy body pathology, significant vascular disease, inflammation, stroke and tumors, as assessed by standard neuropathological examination, and

4. absence of familial AD, as evident by family history.

Patients of all ages fulfilling all the above-stated criteria were included in the study.

Tissue samples were processed following previously described protocols (Zafar *et al.*, 2018). Briefly, whole brains were cut into two parts. One hemisphere was fixed with formalin (4%), followed by treatment with formic acid. After fixation and decontamination, this hemisphere was stored at -80°C until further use for neuropathological examination. Tissue sections of one-centimetre thickness were excised from the other hemisphere of each brain. Dissected tissues were immediately frozen and stored at -80°C until used for biochemical investigations.

For immunohistochemistry, a separate cohort was used. Detailed description is given in annexure data table (Table 13). Samples were obtained from the University Medical Centre Hamburg-Eppendorf. Diagnosis of all these cases was confirmed neuropathologically by a combination of ABC score, Thal staging for amyloid deposition (A), Braak staging for neurofibrillary tangles (NFTs) (B), and the Consortium to Establish a Registry for Alzheimer's Disease (CERAD) neuritic plaque score (C), as recommended by the current criteria of the National Institute on Aging, USA (NIA) for AD. Ethical approval was attained from the ethics committee of the University Medical Centre Göttingen, and all the procedures were followed in accordance to ethics regulations (Nr. 1/11/93 and Nr. 9/6/08).

2.2.1.1 Pathological profiles

Non-demented control cortical samples exhibited mild pathology (Braak stage I – II). Both spAD and rpAD samples had AD pathologies ranging from Braak stage III to VI. The samples for AD subtypes were included without co-pathologies. Likewise, the sCJD subtypes (sCJD-MM1 and -VV2) cohort was exclusively composed of prionopathies. Details of the cohorts are described in the annexure data tables (Tables 11–13). No significant differences were observed in the age distribution and postmortem intervals (PMI) among the disease groups investigated in the study, as shown in Fig. 5 and 6.

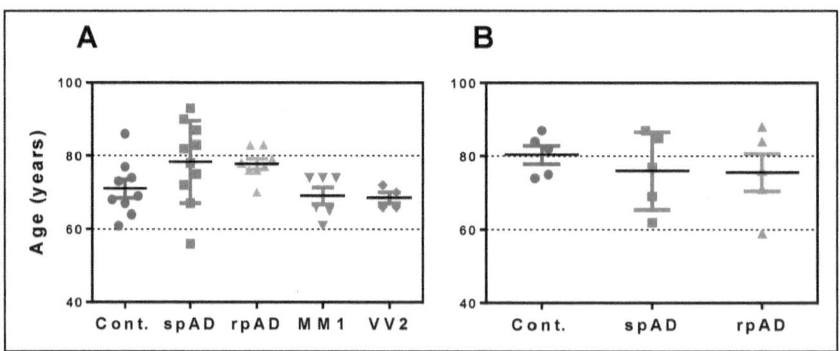

Figure 5: Patient cohorts included in the present study. A) Comparison of ages of the control and diseased subjects used for RBPome isolation, immunoblotting and qRT-PCR analysis. B) Comparison of ages in patient cohort used for immunohistochemical analysis.

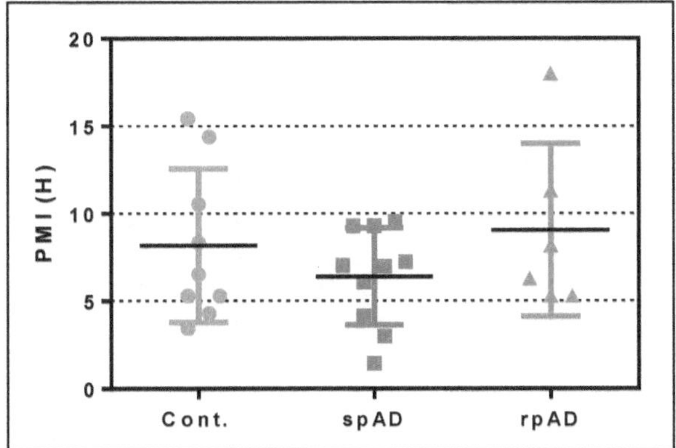

Figure 6: Comparison of postmortem intervals: Comparison of postmortem intervals to the time of autopsies from cases used in the present study.

2.2.2 Molecular biology methods

2.2.2.1 Genomic DNA extraction

Genomic DNA was isolated from human brain frontal cortex samples for *APOE* genotyping. DNA extraction was performed using the purelink genomic DNA isolation kits, in accordance with the manufacturer's protocols. Approximately 30 µg of tissue was used for each extraction. Concentration estimation of DNA was performed with Nanodrop (OD_{260} = 50 µg for dsDNA). Integrity of DNA was confirmed by agarose gel electrophoresis, followed by ethidium bromide staining.

2.2.2.2 *APOE* genotyping

The genotyping of the *APOE* polymorphism was performed using the *APOE* Strip Assay kit (GenoType *APOE*, Hain Lifesciences), as described previously (Al-Asmary *et al.*, 2015). Briefly, the procedure involves two steps: the first involves polymerase chain reaction (PCR) amplification using biotinylate primers, followed by reverse hybridization of the amplified products on a test strip with allele specific oligonucleotides immobilized as an array in line. Biotinylated sequences bound to the strips were detected by streptavidin alkaline phosphatase, in order to develop color with color substrates. The frequency of *APOE* gene polymorphisms is represented for both spAD and rpAD cases in Fig. 7. An increasing trend was observed for alleles 3/3 and 3/4 in rpAD cases.

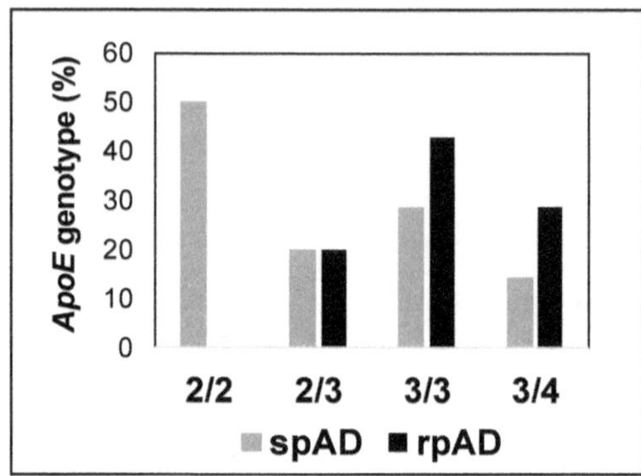

Figure 7: *APOE* **genotype**: The percentage of *APOE* genotype from slow progressive (spAD) and rapidly progressive Alzheimer's disease (rpAD) patients included in the present study.

2.2.2.3 RNA extraction

For the preparation of all reagents for RNA isolation, RNase-free water was used. Before RNA extraction, surfaces and lab-ware were cleaned with RNAseZap (Thermo Fischer Scientific). Total RNA was extracted from the human brain frontal cortices of diseased and control subjects using RNeasy Plus Universal kits (Qiagen, Germany), including DNase treatment in accordance with the manufacturer's instructions. RNA concentration was measured using a Nanodrop 2000 (Thermo Scientific) apparatus. Ratios of A_{260}/A_{280} were also calculated for all samples. RNA integrity was confirmed by using a Bioanalyzer (Agilent Technologies, Santa Clara, CA). Samples bearing RNA integration number (RIN) ≥ 5 were used for further analysis.

2.2.2.4 Tissue lysis for pull-down assay

Frontal cortex tissues from each subject were lysed in 'tissue protein extraction reagent (T-PER; Thermo Fischer Scientific), supplemented with phosphatase and protease-inhibitor cocktails (Roche, Germany). The concentration of isolated proteins was determined using Bradford assay (Bio-Rad). Total protein extract concentration was kept greater than 2 mg/mL, such that there is significant dilution into the binding reaction buffer.

2.2.2.5 RNA pull-down assay

Total brain-derived RNA was end-labeled with desthiobiotin using T4-RNA ligase from Pierce RNA 3'-End Desthiobiotinylation Kit, which was a part of the Pierce magnetic RNA-protein pull-down kits (Thermo Fisher Scientific). The labelling efficiency of experimentally labelled samples was estimated by dot blotting (Thermo Scientific Chemiluminescent Detection Module, Product No. 89880), according to instructions from the manufacturer.

Labelled RNA was used for the enrichment of RBP complexes, according to the manufacturer's recommendations (Thermo Fischer Scientific). Briefly, experimentally labelled RNA was bound to 50 µL of streptavidin magnetic beads in RNA capture buffer and incubated on a rotation wheel for 30 min at room temperature (RT). Beads were washed three times with 20 mM Tris (pH 7.5). Bead-bound RNA was incubated with total protein extract, which was isolated from the human brain frontal cortex, for 1 hour (hr) in RNA-protein binding reaction buffer at 4°C followed by another three washes. Bound protein complexes were eluted with biotin elution buffer and processed for mass spectrometry (MS) analysis. In this assay, streptavidin magnetic beads were mixed with protein extract in the absence of biotinylated transcript as a control for nonspecific binding.

2.2.2.6 Label-free quantification mass spectrometry (LFQ-MS) analysis

Mass spectrometry analysis was carried out as published previously (Zafar *et al.*, 2017). Briefly, isolated protein complexes were separated by 4-20% Bis-Tris gels (NuPAGE Novex Bis-Tris Mini gels, Invitrogen) for a length of ~1 cm followed by Coomassie staining. The bands were excised from the gel into small slices (1-2 mm^2). The gel pieces were initially rinsed with ddH$_2$O followed by reduction (10 mM dithiothreitol [DTT] for 30 min at 56°C) and alkylation (55 mM iodoacetamide [IAA] at RT in dark for 1hr). Then, the gel slices were washed with acetonitrile (ACN) for 15 min and dried using a SpeedVac. The gel slices were incubated overnight with the minimum possible amount of trypsin (12.5 ng/µL in 0.025 M aqueous ammonium bicarbonate) at 37°C. After digestion with trypsin, ddH$_2$O (10 µL) was added to the slices for 15 min at 37°C, followed by the addition of ACN (80 µL) for 15 min at 37°C. Supernatant was isolated after a short-spin. Residual peptides were recovered from slices by incubation for 15 min at 37°C with 5% formic acid (FA) (65 µL). Again, ACN (65 µL) was added for 15 min at 37°C. Supernatant from this step was collected and added to previous supernatant. Total volume of supernatant was evaporated in a vacuum concentrator to dry. These samples were suspended in ACN (30%, 10 uL) and triflouroacetic acid (TFA, 0.1%).

For MS measurements, the peptide mixture was concentrated on a Reversed Phase-C18 precolumn (0.15 mm ID x 20 mm, self-packed with Reprosil-Pur 120 C18-AQ 3 µm material), followed by separation using Reversed Phase-C18 nanoflow chromatography on a Picofrit column, 0.075 mm ID x 200 mm (New Objective, Woburn, USA) and a 15 min linear gradient on an Easy nLC-1000 nanoflow chromatographic system (Thermo Fisher Scientific). Samples were measured on a Q-Exactive hybrid quadrupole/orbitrap MS system operated under Excalibur v2.4 software (Thermo Fisher Scientific).

Analysis was performed by a Top10 method in the Data Dependent Acquisition mode. Tandem mass spectra were extracted for database searching using the Raw2MSM v1.17 software (Max Planck Institute for Biochemistry, Martinsried, Germany). MS/MS samples were analyzed using the Mascot (Matrix Science, London, UK; version 2.5.1) set for searching UniProt/SwissProt database (release 02/17 filtered for *Homo sapiens*, 92928 entries). Mascot was searched with a fragment ion mass tolerance of 0.050 Da and a parent-ion tolerance of 10.0 PPM. Carbamidomethyl of cysteine was specified in Mascot as a fixed modification. Oxidation of methionine, acetylation of the N-terminus and phosphorylation of serine, threonine and tyrosine were specified in Mascot as variable modifications and 2 missed cleavages were allowed. Scaffold (version Scaffold 4.8.4, Proteome Software Inc., Portland/OR, USA) was used to validate MS/MS-based peptide and protein identifications. Peptide identifications were accepted if they could be established at greater than 95% confidence. For protein identifications, a minimum of two peptide counts and a confidence threshold of 99% was used. Protein probabilities were assigned by the Protein Prophet algorithm, as described by Nesvizhskii *et al.* (2003). Proteins sharing similar peptide sequences were classified into defined clusters.

2.2.2.7 Differential enrichment analysis of RBPome

For a detailed analysis of the isolated proteome, three different approaches were used, here labelled approach A, B and C. Approach A was used to identify proteins differentially enriched in the various disease groups. Pairwise *t*-tests (two-sided) were performed for all disease group combinations using Perseus software (version 1.5.0.31) (Tyanova and Cox, 2018), with a p-value < 0.05 and multiple testing correction using the Benjamini-Hochberg (BH) method. Using label-free MS to identify and quantify proteins, zero values were observed in the data for several proteins. Missing or zero values appeared when the mass spectrometer could not detect peptides having abundances below the censoring cutoff of the mass spectrometer. Such values were informative because they were below the lowest abundance observed for a peptide. In such situations, when quantitative values were missing in one group but were present in other groups, this more likely represented differences in abundance between groups, which might indicate interesting features specific to that group. Therefore, zero values were imputed by half of the minimum value of total spectrum count values, to have statistical analysis of the proteomic candidates (Approach A). Fold change (FC) for all comparisons' thresholds was set at ±1.5 and a p-value < 0.05 for significance. Proteins which were identified as significantly enriched were used to make heatmaps using Perseus software (version 1.5.0.31). Volcano plots were also calculated using Perseus software, where FC was $\log_2$ transformed, so that the data were centered on zero, while the p-values were transformed into $-\log_{10}$.

In addition, approach B was used. Given the exploratory nature of this discovery-based proteomic work-flow, the criteria were relaxed and proteins with a single quantitative value were included in the analysis to compare all the disease groups. Venn diagrams were prepared, showing similar and unique proteins with functional enrichment and analysis tool (FunRich).

To have a deeper proteome coverage of isolated RNA-binding proteome of particularly rapidly progressive forms of dementia (rpAD, sCJD-MM1 and sCJD-VV2 subtypes), we used SWATH-MS quantitative proteomic analysis as the approach C, which also served as a confirmatory method for the LFQ-MS analysis. Detailed description of SWATH-MS is described in Section 2.2.15.8. Quantified proteins were analysed using Perseus software (version 1.5.0.31), to identify significantly modulated proteins between two pairs of comparisons, namely rpAD vs sCJD-MM1 and rpAD vs sCJD-VV2. Significantly abundant proteins were visualized by heatmaps generated using Perseus software (version 1.5.0.31).

2.2.2.8 Quantitative real-time PCR (qRT-PCR)

Complementary DNA (cDNA) was synthesized from brain-derived total RNA (1 µg), using the High Capacity cDNA Reverse Transcription Kit (Thermo Fisher Scientific). The synthesis was performed according to the instructions from the kit protocol. The resulting cDNA was diluted (1:10) and stored at -20°C. All primer pairs for qRT-PCR assays were designed with Primer 3. The product size of all the primers was in a range of 100 to 150 base pair long. The primer sequences were tested using NCBI/Primer BLAST and synthesized by Eurofins Genomics (Table 14). Reaction mixtures for RT-PCR were prepared by mixing 1 uL of 10x PCR reaction buffer (Roche), 1 uL of dimethylsulphoxide (DMSO) (Sigma), 0.5 uL of 1:1000 dilutions of SYBR Green (Sigma), 0.2 mmol of each dNTP (Roche), 0.15 units of Taq-polymerase, 10 pm/uL of forward and reverse primers and 1 uL of cDNA (1:10 dilut-ed); the volume was filled up to 20 uL with RNAse-free water. Amplification was per-formed using Light Cycler 480 (Roche), with an initial denaturation at 95°C for 2 min, followed by 40-PCR cycles (denaturation at 95°C for 30 seconds [sec], annealing at 56°C for 30 sec, and extension at 72°C for 30 sec). Reactions were done in tripli-cates. No template controls were used to ensure reaction specificity. The data were analysed with Light Cycler 480 software SW1.5.1 (Roche) and values were normal-ised to *GAPDH*. The comparative C_t method ($2^{-\Delta\Delta Ct}$) was used to measure relative fold change mRNA expression (Livak and Schmittgen, 2001).

2.2.2.9 Preparation of brain homogenates for protein analysis

Frontal cortex tissues were lysed from spAD, rpAD, sCJD (MM1 and VV2 subtypes) and non-demented control subjects, along with cortical tissues from mouse brains, as published previously (Zafar *et al.*, 2018). Briefly, brain tissues were homogenized (10%, wt/vol) using tissue lyser LT (Qiagen) in ice-cold tissue lysis buffer containing protease and phosphatase inhibitors (Roche). To remove insoluble debris, lysates were centrifuged at 14000 rpm for 30 min at 4°C. Protein quantification of isolated extracts was performed by Bradford protein assay. Bradford assay dye reagent was diluted with ddH2O at a ratio of 1:5. Dilutions of protein standard, bovine serum albumin (BSA) were prepared (0.0–1.0 mg/mL) in ddH$_2$O. Total protein extracts from experimental samples were diluted at a ratio of 1:20. From this dilution, 20 µL was mixed with 980 uL of Bradford working solution and the mixture was incubated for 10 min at RT. All dilutions of protein standard and experimental samples were measured at a wavelength of 595 nm. Measured optical density values were used for calculation of protein concentration.

2.2.2.10 Immunoblotting

Equal amounts of proteins (30–50 ug) from each brain lysate or cell lysate were resolved by molecular weight on 12% SDS-PAGE polyacrylamide gels (prepared in house) or 4–12% Bis-Tris gels (NuPAGE 4–12% Bis-Tris Protein Gels, Invitrogen). The protein marker (Precision plus protein dual color standards from Bio-Rad) was used to visualize the correct separation of the proteins and to confirm the correct protein band sizes. After separation on the gel, proteins were transferred onto a polyvinylidene difluoride (PVDF) membrane with a 0.45 µm pore size using a semi-dry blot chamber (Bio-Rad, Hercules, USA) for 1 hr. The membranes were blocked for 1 hr at RT in blocking reagent (5% non-fat dry milk in TBS-T or PBS-T), followed by incubation with primary antibodies at 4°C overnight. All the primary antibodies (total tau, phospho-tau (S199), SFPQ, TIA-1, VCP, β-actin, BRD-4, and GAPDH) were diluted in blocking buffer. The dilutions of all the primary antibodies are given in Table 2. Next, the membranes were washed in PBS-T/TBS-T and incubated with secondary antibodies coupled to horseradish peroxidase (HRP) for 60 min at RT. Protein bands were detected using the enhanced chemiluminescent (ECL) method with Chemi-Doc (Bio-Rad). The densitometric analysis was performed with Image Lab software (3.0.1).

2.2.2.11 Immunohistochemistry

Formalin-fixed and paraffin-embedded human brain cortical sections (5 µm thick) from patients with spAD and rpAD as well as non-demented controls were analysed by co-immunofluorescence using protocols validated previously (Krasemann *et al.*, 2017; Vanderweyde *et el.*, 2012). Briefly, tissue sections were deparaffinized using xylene, rehydrated and then proceeded for antigen retrieval with citrate buffer (pH 6.0) for 1 hr at 95°C. The slides were cooled for 30 min at RT and then washed with PBS for 1 hr with repeated changes. Next, the sections were permeabilized with

0.2% Triton X-100 for 10 min, followed by two washes with PBS. The slides were then exposed to UV-A and UV-B for 20 min, to remove background fluorescence. The sections were then blocked with a blocking reagent (1% BSA, 10% FBS in PBS) for 1 hr at RT, treated with Sudan black to remove lipofuscin fluorescence and then washed with PBS for 30 min. The primary antibodies specific for SFPQ, TIA-1, phospho-tau (S199) and oligomeric tau (T22) were incubated overnight at 4°C. Dilutions of all the primary antibodies are given in Table 2. After incubation, the slides were washed in PBS for 1 hr. The secondary antibodies, Alexa488, Alexa546 and Alexa555 (Invitrogen), were incubated for 2 hrs at RT followed by counter staining with TO-PRO-3 iodide for 1 min to visualize cell nuclei. After washing with PBS, the sections were mounted with Fluoromount-G (Thermo Fischer Scientific). After secondary antibody incubation, the whole procedure was carried out in dark.

2.2.2.12 Confocal laser scanning and image analysis

Imaging was performed with the confocal laser scanning microscope TCS-SPE from Leica, using 543 and 633 nm helium-neon and 488 nm argon excitation wavelengths. All images were separately analysed for co-localization using ImageJ (WCIF plugin) and FIJI software. Threshold Mander's overlap coefficient (tM) and Pearson's linear correlation coefficient (rP) were calculated to measure the strength and direction of linear correlations between two fluorescence channels. In addition, intensity correlation analysis (ICA) was performed. This analysis generated scatter-plots of channel A and channel B against the product of the difference of each pixel A and B intensities from their respective means (PDM) (Li *et al.*, 2004). The resulting plots emphasized the high intensity stained pixels and allowed to identify protein pairs that vary in synchrony (positive PDM values), randomly (around 0), or independently (negative PDM values) within the cell.

2.2.3 Microbiological methods

2.2.3.1 Culturing and storage of *E. coli*

Plasmids were purchased from Addgene in the *E.coli* strain DH5α in agar stab and were propagated in-house. For long-term storage, bacterial strains were mixed with an equal volume of glycerol (50%) and stored at -80ºC.

2.2.3.2 Extraction of plasmid DNA

For the extraction of plasmid DNA, Hispeed Plasmid Midi Kit (Qiagen) was used. A single bacterial colony was inoculated into 5 mL of LB medium containing ampicillin (100 µg/mL) followed by incubation for 12-16 hrs with continuous shaking (180–250 rpm) at 37°C. The resultant bacterial pellets were harvested by centrifugation at 5,000xg for 10 min at 4°C. All further procedures were performed according to the instructions of the kit protocol.

2.2.4 Cell biology methods

2.2.4.1 Cryopreservation of mammalian cell lines

In order to be stored for longer, the cells were preserved in liquid nitrogen in DMSO. Cells at confluency (70–90%) were harvested and centrifuged at 400×g for 5 min. The cell pellet was resuspended in freezing medium (5% DMSO in culture medium) and immediately transferred to 1 mL cryogenic vials, followed by overnight freezing at -80°C. Finally, the vials were transferred to liquid nitrogen.

2.2.4.2 Cell culturing and maintenance

The cells were taken from the liquid nitrogen tank and incubated for 2 min in a 37°C water bath until nearly 80% thawed. The cells were then gently mixed with 10 mL of pre-warmed culture medium (DMEM, 10% FBS, 1% PS) and centrifuged at 400xg for 5 min. The supernatant was removed and the pellet was suspended in 14 mL of the culture medium. The cells were cultured in flasks and plates at 37°C, 95% humidity and 5% CO_2. After 24 hrs, the media was replaced with a fresh 14 mL culture medium. Once the cells reached confluency, they were washed once with PBS followed by trypsinization using 2-3 mL of trypsin/EDTA solution. After trypsinization, cells were centrifuged at 400xg for 5 min and then washed with PBS. Cell counting was performed using a haemocytometer (0.1 mm sample depth). The cell pellet was resuspended in a fresh pre-warmed culture medium and cells were grown to approximately 70–90% confluency. All cell lines were cultured between 5 and 20 passages.

2.2.4.3 Stress model for stress induction

Confluent cells (HeLa and SH-SY5Y) were subjected to oxidative stress by adding a pre-warmed culture medium containing sodium arsenite (0.6 mM) for 1 hr at 37°C.

After 1 hr of treatment, the cells were either fixed with 4% paraformaldehyde (PFA) for immunocytochemistry or processed for immunoblotting analysis.

2.2.4.4 Immunocytochemistry

The cells were grown in T75 flasks with culture medium (DMEM, 10% FBS, 1% PS). At confluency (70–90%), cells were trypsinized and seeded (5 x 10^4) on glass coverslips (13 mm) in 24-well plates. Following the stress treatment as described above, the cells were fixed with 4% PFA for 20 min at RT, followed by three washes with ice-cold PBS for 5 min each. Permeabilization was achieved using 0.2% Triton X-100 in PBS for 10 min, followed by three washes with PBS for 5 min each. To avoid non-specific binding, cells were incubated with blocking buffer (1% BSA, 10% FBS in PBS) for 30 min at RT. The primary antibody and, in case of double labelling both primary antibodies, were diluted in 1% BSA in PBS, followed by overnight incubation at 4°C. The cells were washed three times with PBS (each wash was for 5 min) followed by incubation with secondary antibodies (see Table 3) diluted in 1% BSA in PBS for 2 hrs at RT in the dark. The cells were rinsed three times with PBS for 5 min each in the dark to remove non-specific immunoreactivity. The cells were then counterstained with DAPI for 1 min or a RedDot 2 Far red nuclear stain for 20 min. After nuclear staining, the cells were washed three times with PBS, followed by mounting with one drop of mounting medium (immuo-mount, Thermo Fisher Scientific). The slides were stored in the dark at -20°C or +4°C until they were imaged. The average number of stress granules in each cell was calculated using FIJI software.

2.2.4.5 Subcellular fractionation

Cellular fractionation after stress treatment was carried out as described previously (Suzuki *et al.*, 2010). Briefly, confluent HeLa cells (70–90%) were washed once with ice-cold PBS, scraped and resuspended in 1 mL of PBS. After a short spin (10 sec), the supernatant was removed and the cell pellet was resuspended in 900 µL of the lysis buffer (0.1% NP40 in PBS) (Calbiochem, CA, USA) followed by trituration (5x) using a p1000 micropipette. From this step, 300 uL of the whole cell lysate fraction were isolated and mixed with 4x Laemmli buffer (100 µL). The remaining (600 µL) lysate was centrifuged for 10 sec. The supernatant (300 µL) from this step was separated as the "cytosolic fraction" and mixed with 4x Laemmli buffer (100 µL) followed by boiling for one min. The remaining supernatant in the original tube was removed and the pellet was resuspended in 1 mL of lysis buffer and centrifuged for 10 sec. This supernatant was removed and the pellet (~20 µL) was resuspended in 180 µL of 1x Laemmli buffer and saved as the "nuclear fraction". Whole-cell nuclear fractions containing DNA were sonicated in an ultrasonicator (Level 2, twice for 5 sec), followed by boiling for 1 min. From each fraction, 10 µL were proceeded for immunoblotting analysis with primary antibodies specific for total tau (tau-5), phospho-tau (S199), TIA-1, SFPQ, β-actin, BRD4, and GAPDH.

2.2.4.6 Cell lysis and protein extraction

Total protein extracts were prepared from 70-90% confluent HeLa and SH-SY5Y cells. After stress treatment, all the cell lines were washed with 1x PBS, scraped and resuspended in lysis buffer (50 mM Tris-HCl, pH 8, 1% Triton X-100, 0.5% CHAPS, 1mM DTT, protease, and phosphatase inhibitors). Lysates were sonicated using an ultrasonicator on ice, followed by incubation for 1 hr at 4°C with shaking. The lysates were centrifuged at 14000 rpm for 30 min at 4°C. The supernatants were transferred to new tubes and proteins were quantified as described above. To harvest cells after transient transfections, cells were trypsinized and washed with PBS followed by centrifugation at 400xg for 5 min at 4°C. The washed cell pellets were lysed, and protein quantification was performed as above.

2.2.4.7 Tau transfections

Plasmids for human wild-type tau (pRK5-EGFP-tau) and mutated tau (pRK5-EGFP-tau P301L) were obtained from Addgene. The plasmids were propagated in *E. coli* DH5α strain and extracted from the the transformed bacteria using the Hispeed Plasmid Midi kit (Qiagen). Transient transfections were achieved using the Lipofectamine 2000 (Invitrogen) reagent, according to the instructions of the manufacturer. HeLa cells (2x 10^5/well) were plated in 6-well plates and cultured for 24 hrs in culture medium. The cells were washed with Opti-MEM I, followed by transfection with 2 µg of DNA/well in the Opti-MEM I medium. The cells were collected from cultures after 24 and 48 hrs post-transfection.

2.2.4.8 SWATH-MS for global proteomics

2.4.8.1 Library preparation

Analytical-grade reagents were used for protein extraction and digestion. Sterile water (Ampuwa, Fresenius, Bad Homberg, Germany) was used to prepare all buffers and solutions. For the preparation of the spectral peptide library, digested protein extracts (normalized for protein amounts) from each sample were pooled to a total amount of 220 µg and separated into fourteen staggered pooled fractions, using an Äkta pure (GE Healthcare) with a Hypersil Gold C18 column (diameter 150x2,1 mm, Particle size: 3 µm). Protein digests were analyzed on an Eksigent nanoLC425 nanoflow chromatography system associated with TripleTOF 5600[+], a hybrid triple quadrupole TOF mass spectrometer with a Nanospray III ionization source (ionspray voltage 2400V, Interface heater temperature 150°C, Sheath gas setting 12). The peptides were dissolved in loading buffer (0.1% formic acid, 2% acetonitrile in optima water [Thermo Scientific]) to a final concentration of 0.3 µg/µL and spiked with a synthetic peptide standard used for retention-time alignment (iRT Standard, Schlieren, Schweiz). For every measurement, digested proteins (1.5 µg) were enriched on a precolumn (PharmFluidics µPAC Trapping Column) and separated on a PharmaFluidics µPAC microchip-based separation analytical column with 50 cm length, using a 120 min linear gradient of 5%–45% ACN, 0.1% FA at a flow rate of 300 nL/min. Qualitative LC-MS/MS measurement was carried out using a Top 20 data-dependent acquisition (DDA) mode with a mass range of m/z 350–1250 for 250 milliseconds (ms). The resolution for data extraction was 30,000 full width at half maximum (FWHM). MS/MS scans of mass range m/z 180–1600 at a resolution of 17,500 FWHM for 85 ms and a precursor isolation width of 0.7 FWHM with a cycle time of 2.9 sec were used. For MS/MS, the criteria for precursor ions selection were as follows: a threshold intensity of more than 125 cps with charge states of 2+, 3+, and 4+, and the dynamic exclusion of 30 sec. In every reversed-phase fraction, two technical replicates were analyzed for the construction of the spectral library.

2.4.8.2 Quantitative SWATH measurement

For the SWATH measurement, MS/MS data were obtained using windows of 65 variable sizes across a mass range of 400–900 m/z. For each biological sample, two technical replicate injections were attained. Protein identifications were attained using ProteinPilot Software version 5.0 build 4769 (AB Sciex) at "thorough" settings. Spectra from MS/MS were searched in the UniProtKB using *Homo sapiens* as reference proteome (revision 04/2018, 93661 entries) at a critical false discovery rate (FDR) of 1%. Spectral library and SWATH peak extraction were attained in PeakView Software version 2.1 build 11041 (AB Sciex). Following retention time correction using the iRT standard, peak areas were extracted using information from the MS/MS library at 1% FDR. The peak areas were summed up to peptide and finally corresponding protein area values that were used for downstream statistical and functional analysis.

2.4.8.3 Gene Ontology analysis and functional network mapping

To gain functional insights from proteomics data, three different enrichment strategies were used. Functional enrichment analysis was initially performed using Perseus software for significantly enriched Gene Ontology (GO) processes, including the Biological Process and Molecular Function by Fisher's exact test. Then, overrepresentation enrichment analysis was performed using the web-based Gene SeT Analysis Toolkit (WebGestalt), in the domains of Biological and Molecular Functions to have a GO Slim summary of enriched terms. Finally, an Ingenuity Pathway Analysis (IPA, Qiagen, USA) was performed to find out canonical pathways associated with up- and down-regulated proteins after tau expression, as described below.

2.4.8.4 Ingenuity Pathway Analysis (IPA)

The proteomic dataset containing the fold change and p-values of significantly regulated proteins was uploaded to IPA for core analysis (Qiagen, USA). The protein candidates from the submitted dataset generated top molecular networks based on Molecular and Biological Functions including canonical pathways, potential upstream regulators, and disease-based networks. The settings for analysis were based on direct and indirect relationships between differentially expressed proteins (DEPs), and these were supported by experimentally reported data from human, mouse and rat studies (Everts *et al.*, 2014). Potential upstream regulators were designated as inhibited or activated, according to the fold change and p-values ($-\log_{10}$-p-values) (Sardiu *et al.*, 2009) of the DEPs. In order to understand the complex relationships among significant disease-based networks, Cytoscape (3.6.1) was used to construct and visualize the networks.

2.2.5 Biochemical methods

2.2.5.1 MTS assay

Cells were grown in T75 flasks with culture medium (DMEM, 10% FBS, 1% PS). At confluency (70–90%), cells were trypsinized and then washed once with PBS. The cells were seeded (10 x 10^3) in 96-well plates and incubated for 18–24 hrs at 37°C. Transfections were performed as described above with plasmids coding for wild-type tau and P301L-tau for variable periods of time (24 and 48 hrs). To measure cell viability, MTS assay (ab197010, Abcam) was used according to the manufacturer's protocol. The culture media was replaced with a fresh medium before treatment with the 3-(4,5-dimethylthiazol-2-yl)-5-(3-carboxymethoxyphenyl)-2-(4-sulfophenyl)-2H tetrazolium MTS reagent. To estimate the effect of tau expression on cell viability, the reduced MTS tetrazolium complex (colored formazan product) was measured. This conversion is a property of metabolically active cells. For color development, cells were incubated at 37°C for 1 hr and the absorbance measurement was taken at 490 nm using a Perkin Elmer Wallac 1420 Victor microplate reader (GMI, USA). Absorbance from the control wells (background) was subtracted from the sample wells.

2.2.5.2 Trypan blue exclusion assay

Cell viability was also assessed through the trypan blue exclusion dye test. Briefly, the cell suspension in the culture medium (25 µL) after trypsinization was mixed with 25 µL of 0.4% trypan blue. After mixing, the 10 uL of cell suspension was loaded onto haemocytometer. The viable (colorless) and dead (blue) cells were counted in each large square of the haemocytometer using a 40x objective for both untreated (control) and sodium-arsenite-treated (stress) cells.

2.2.6 Animal time course and sample collection

All animal experiments were conducted according to the ethical standards of the Regierungspräsidium Tübingen (Regional Council) experimental no. FLI 231/07 file reference number 35/9185.81-2. All procedures were carried out in accordance with the institutional and French national guidelines within the European Community Council Directive 86/609/EEC. The experimental procedures approved by the INRA Toulouse/ENVT ethics committee were used.

The mice model (3xTg-AD) was initially generated and characterized by Oddo et al. (2003). Transgenic mice between 3–4 months of age were inoculated with 10% brain homogenates in PBS from AD-patients, in the thalamus. Both, inoculated and non-inoculated control animals were sacrificed at 4 time-points corresponding to early pre-symptomatic (3 months post-inoculation [mpi]), late pre-symptomatic (6 mpi), early symptomatic (9 mpi) and late symptomatic (12 mpi) stages of the disease, and the brains were collected. All mice (n = 48) were anesthetized and decapitated. The cortical tissues were obtained and immediately stored in liquid nitrogen for further analysis. A graphical representation of the mice model indicating different time points of the sample collection is given in Fig. 8.

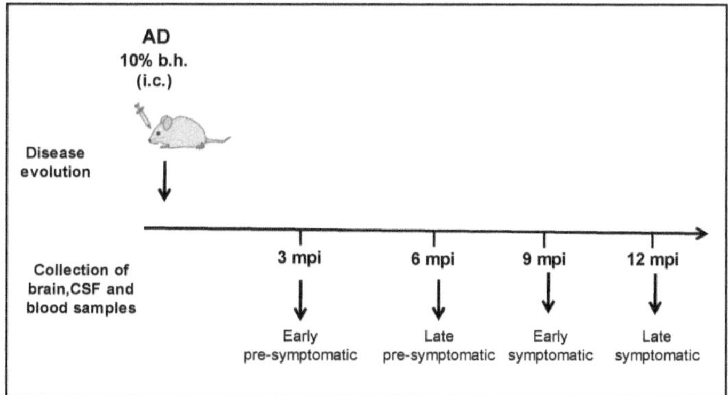

Figure 8: Graphical representation of animal experimentation using the mice model of Alzheimer's disease. Transgenic mice model of AD (3xTg-AD) was used for inoculation. Brain homogenates (b.h.) (10%) from AD patients were injected into thalamus. The samples were collected for early pre-symptomatic (3 months post-inoculation [mpi]), late pre-symptomatic (6 mpi), early symptomatic (9 mpi) and late symptomatic (12 mpi) stages of the disease.

2.2.7 Statistical analysis

All the data in the present study was obtained after performing at least three independent experiments. All the results are described as mean ± SEM (standard error of the mean). The densitometry analysis of immunoblots was performed with Image Lab software (version 3.0.1). Statistical tests were applied using GraphPad Prism 6.01. The data from mass spectrometry analysis were analysed using the Perseus software with adjustments for multiple testing. Hierarchical clustering analysis was also performed with Perseus software. Two-dimensional interaction plots were plotted in R (version 3.4.3), followed by editing using Inkscape (version 0.92). For comparisons between two groups, the Student's *t*-test or Welch's *t*-test was used. For comparisons between three or more groups, a one-way ANOVA followed by the Tukey post-hoc analysis was used. Statistical significance was considered for a p-value < 0.05.

3 Results

Emerging evidence has demonstrated that numerous neurodegenerative disorders including AD display features of RBP pathologies, highlighting a vital role of RBPs in neurodegeneration (Cookson, 2017; Vanderweyde *et al.,* 2016). The study of RBP biology is at its early stages, particularly in Alzheimer's and prion diseases. To this end, we identified and characterized the RNA-binding proteome variations in Alzheimer's and prion diseases.

This chapter is divided into four major sections. The first section describes the identification and characterization of RNA-binding proteomes from postmortem brains of diseased and healthy subjects. Target candidates were prioritized from proteomic analyses for further investigation. The second section is focussed on pathological characterization of target RBPs, particularly SFPQ, in the human brain. The third section describes the cellular model of stress and the tau-pathology model, and, in addition, explores the mechanistic link of SFPQ in disease pathogenesis and progression. This chapter links SFPQ in the human brain to stress granules and tau protein. Finally, in the fourth section, SFPQ and associated proteomic signatures are studied at pre-symptomatic and symptomatic stages of the disease in the 3xTg-AD mice model, in order to uncover early changes during the disease progression (Fig. 9).

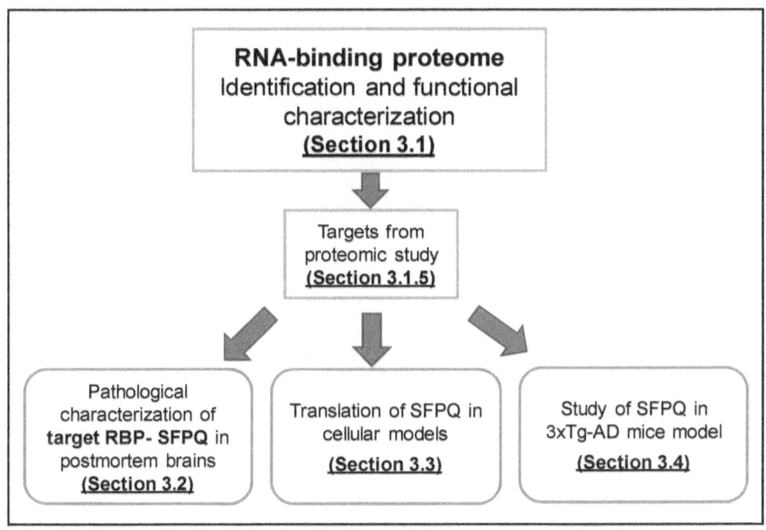

Figure 9: Overview of the results: In the first part of the study, RNA-binding proteins were identified and characterized in postmortem brains from diseased and healthy subjects (Section 3.1). Targets were prioritized for further investigation from the proteomic study (Section 3.1.5). One of these target RBPs, the RNA-binding protein SFPQ (splicing factor proline and glutamine rich), was pathologically characterized in the human brain (Section 3.2). Then, these target candidates were translated in the cellular (Section 3.3) and animal models (Section 3.4) to investigate mechanistic links with pathological features of the disease.

3.1 RNA pull-down assay and mass spectrometry analysis of RNA-binding proteome

In this study, we utilized brain-derived, total RNA to pull down RBPome complexes from the human brain of spAD, rpAD, sCJD as well as control subjects. Mass spectrometry analysis was performed to identify and quantify isolated RNA-binding protein complexes. A schematic outline of the RNA pull-down/MS procedure used in this study is shown in Fig. 10A.

Total RNA was isolated from the frontal cortical region of the brain of diseased and control subjects. Brain-isolated RNA was labelled with desthiobiotin at 3'-end to minimize any disturbance to the secondary structure of the RNA. Desthio-biotinylated-RNA was bound to streptavidin magnetic beads and oriented to enrich RNA-binding protein complexes. Isolated protein complexes were identified by two MS approaches: global proteomic approach (spectral counting-based, label-free quantification [LFQ]) and a variant of data-independent quantitative MS (Sequential Window Acquisition of All Theoretical Mass Spectra [SWATH-MS]) (Fig. 10A).

After the identification and quantification of RBPome candidates, a combination of bioinformatic, computational, and biochemical approaches was used to comprehensively analyse the isolated RBPome (Fig. 10B).

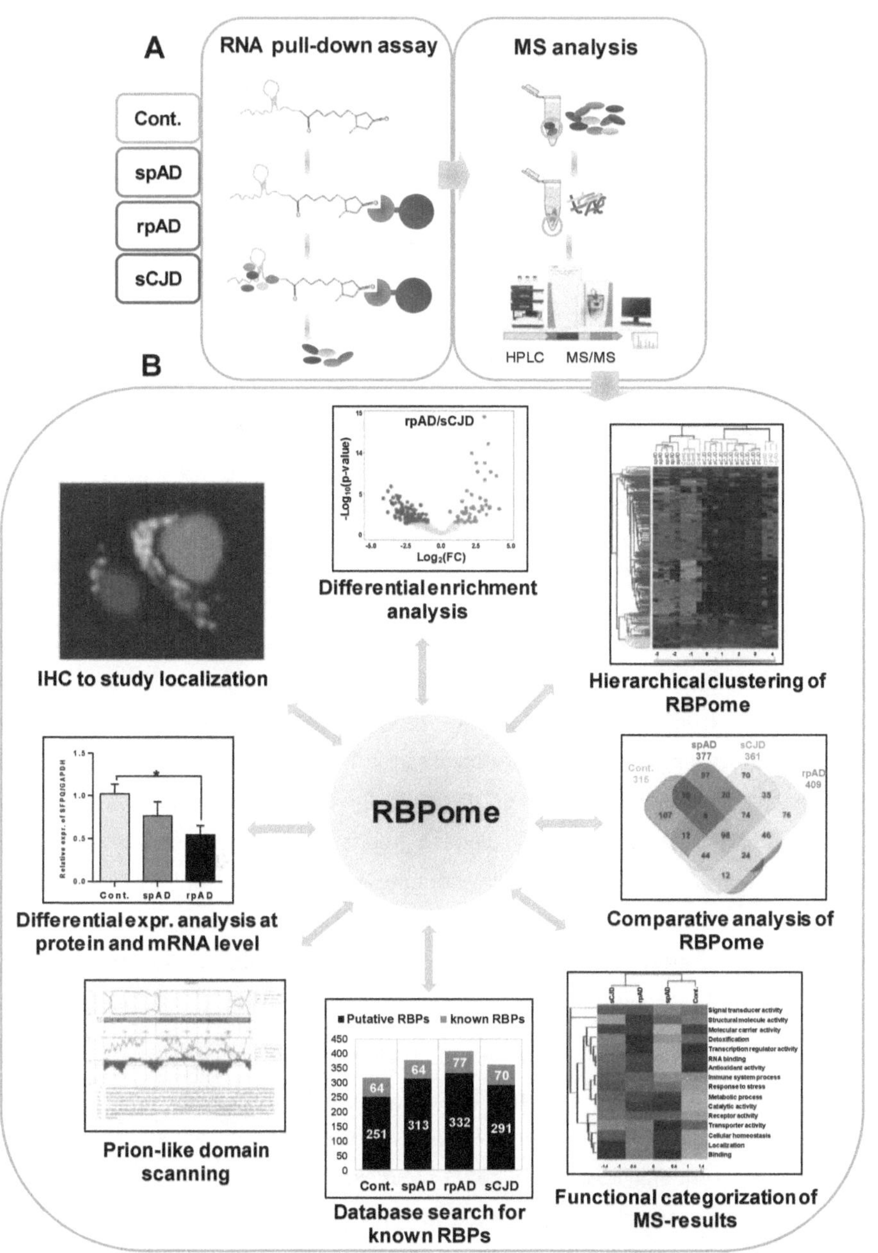

Figure 10: Identification of RNA-binding proteome by total RNA pull-down assay coupled with mass spectrometry analysis. A) Total RNA was isolated from the frontal cortical region of 20 diseased and control cases. Brain-derived RNA was desthiobiotinylated at 3'-end to minimize disturbance to the secondary structure of

RNA. Desthiobiotinylated RNA was bound with streptavidin magnetic beads. Bead-bound RNA was oriented to pull down RBP complexes from total tissue lysates of corresponding diseased and healthy subjects. Bead-only controls were used for non-specific binding. Isolated RBP complexes were digested into peptides followed by identification and quantification by mass spectrometry. In total, 1,091 proteins were detected and quantified at a minimum peptide count of 2 and a protein confidence threshold of 99%. **B) Target selection from proteomic investigation and their pathological characterization in the postmortem brains**. A combination of bioinformatic and computational approaches was used to identify significant hits from the proteomic study, including differential enrichment analysis of MS data, hierarchical clustering analysis to visualize global proteome profile, comparative RBPome analysis for similar and unique proteins, Gene Ontology (GO) functional enrichment analysis, database search for identification of *bona-fide* and novel or putative RBP candidates, and prion-like domain scanning with the PLAAC database. Target candidates from proteomic investigation were pathologically characterized in the postmortem human brain, using various techniques including immunoblotting, qRT-PCR, and immunohistochemical analysis.

3.1.1 Global enrichment profile of RNA-binding proteome

Label-free quantitative mass spectrometry analysis identified a total of 1,091 proteins, with minimum peptide count of 2 and a protein confidence threshold of 99%. For detailed analysis of the identified proteome, three strategies were adopted (Fig. 11). A detailed description of all the approaches (A, B and C) can be found in "Methods Section". In approach A and B, two subtypes of sCJD (MM1 and VV2) were merged, and only proteins that were common in both subtypes were considered to be a general representation of the sCJD group. Firstly, to gain insight into the global similarities and differences between disease groups, RBP-candidates were uploaded to Perseus software, followed by pairwise comparisons between all the group combinations using *t*-test (approach A). The proteins having a p-value < 0.05 and FC > ±1.5 were considered as significantly abundant between any of the experimental groups (Fig. 12A).

Using significantly enriched proteins in the hierarchical clustering analysis, four different expression signatures were found, which are represented in the form of heatmap (Fig. 12B). Columns of the heatmap represent all the disease groups, while the rows represent significantly abundant proteins. Clustering of the biological replicates across the vertical axis generated two main column clusters. Based on RBP signatures, rpAD and sCJD groups appeared largely similar as they were segregated into a single cluster and differed from the control group (Fig. 12B). The control and spAD cases were segregated into a separate cluster based on the full set of differentially enriched proteins, which indicated similarities in the protein profiles between control and spAD groups.

Overall, clustering of the individual replicates indicated low biological variance and segregation of all the groups, showing the reproducibility and robustness of the work-flow (Fig. 12B).

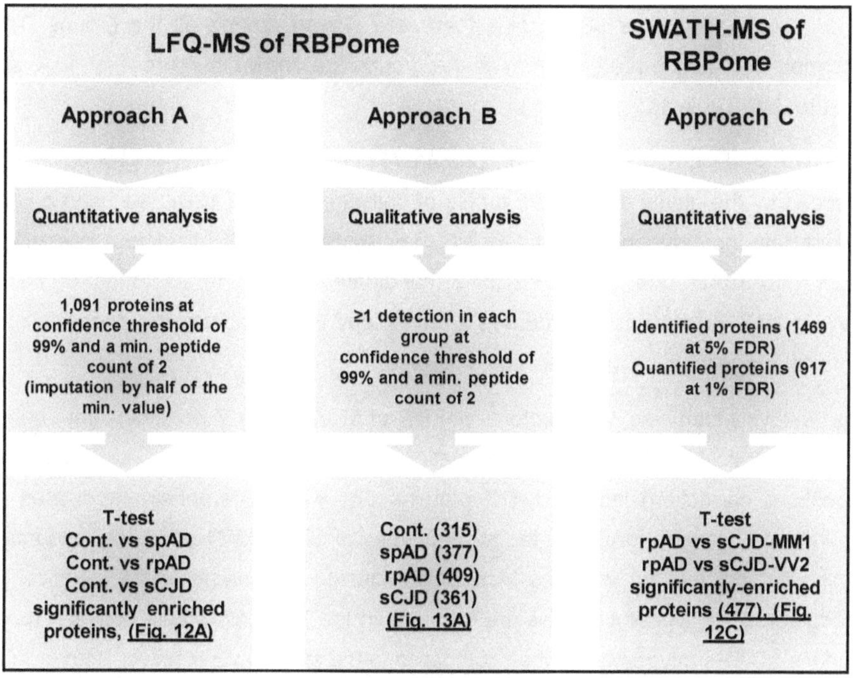

Figure 11: **Statistical analysis of RBPome candidates**: Two different strategies (approaches A and B) were used to obtain a more detailed analysis of RBPome candidates identified from LFQ-MS. **Approach A**: Statistical analysis of isolated proteomic candidates by Perseus software. Zero values from the total spectral counts of the proteins were imputed by half of the minimum value followed by pairwise Student's *t*-test comparisons between all the group combinations, to find out significantly enriched proteins in each group. **Approach B**: Qualitative analysis of the RBPome. Proteins with a single quantitative value were included in this approach from all the groups to have a broader impression of the isolated proteome. In order to find out common and unique proteins among all the groups, a comparative RNA-binding proteome profile was obtained, based on specifically identified proteins in each group. **Approach C**: SWATH-MS of rpAD and sCJD (MM1 and VV2 subtypes) followed by pairwise comparisons, as above. The numbers of resulted proteins, and selected proteins is described in figures as indicated.

Given the exploratory nature of this discovery-phase proteomics workflow, we re-laxed the criteria and included all the proteins with one or more than a single-detection in each group, due to the high sensitivity of the MS analyser (approach B) (Fig. 11). The proteins which were specifically enriched in each group were further analysed to have a comparative proteome profile of the RBP-candidates. This analysis revealed disease-subtype specific differences in the enriched RBP candidates (Fig. 13A). There were 98 proteins that were shared among all the groups. The common and unique proteins from all the groups are shown in figure (Fig. 13A, annexure data Table 15).

The RBPome variations were highly prominent in the sCJD and rpAD groups, as evidenced by the global enrichment profile of both groups (Fig. 12B). We used a variant of data-independent quantitative MS, namely SWATH-MS, to gain a deeper insight into rpAD and sCJD RBPome candidates (Fig. 11, approach C). The SWATH-MS variant has emerged as a technology that combines the deeper proteome coverage with the quantification accuracy.

This analysis identified 1469 proteins at FDR of 5%, while 917 proteins were quantified at a critical FDR of 1%. Bioinformatic analysis (Welch's t-test with Benjamini-Hochburg correction) indicated 477 proteins that were differentially modulated in abundance between rpAD and the two subtypes of sCJD, MM1 and VV2. One interesting finding, evident from the hierarchical clustering analysis of the significantly modulated RBP signatures, was the similarity in the RBPome profile between rpAD and the sCJD-subtype MM1, when compared to the sCJD-VV2 subtype. Both groups (rpAD and sCJD-MM1) were segregated into a single cluster, whereas the sCJD-VV2 subtype was segregated into a separate cluster (Fig. 12C).

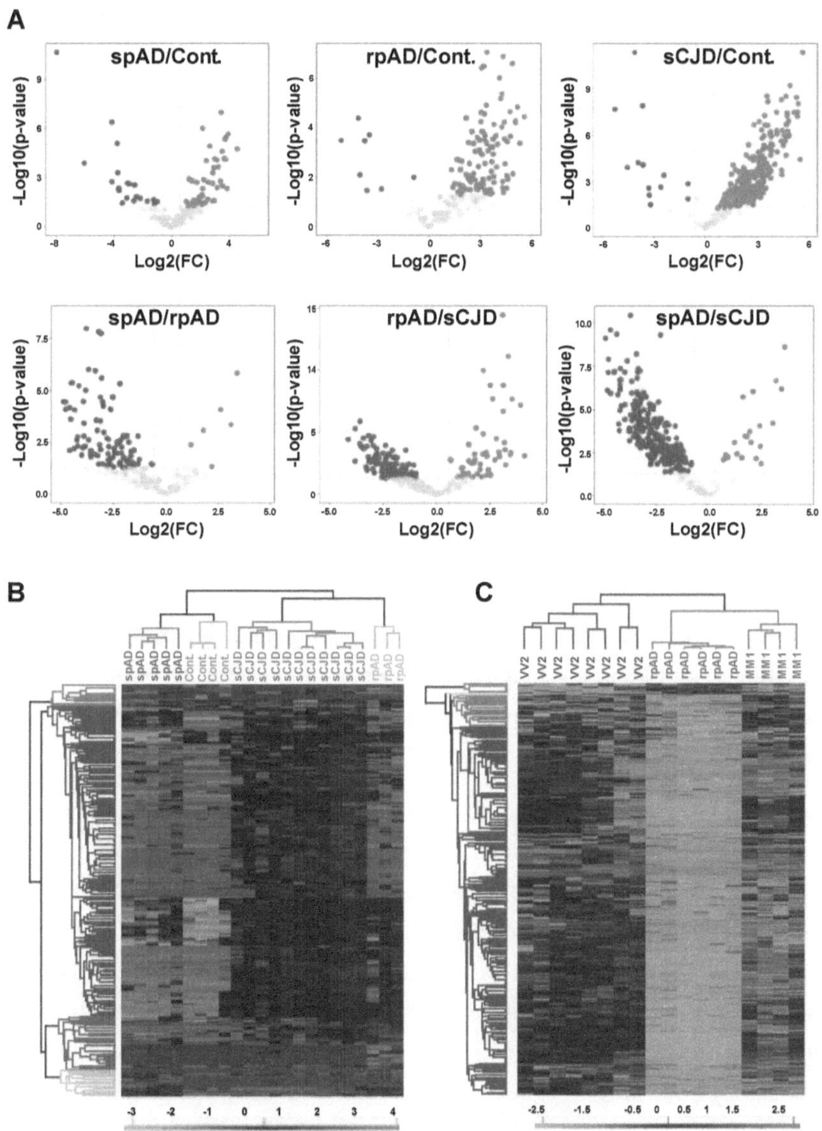

Figure 12: Global enrichment profile of RBPome candidates: **A)** Volcano plots of pairwise comparisons displaying, the -log₁₀-p-values (y-axis) and the log₂(FC) of the proteins that were significantly abundant (x-axis) in all the group combinations (spAD vs control, rpAD vs control, sCJD vs control, spAD vs rpAD, rpAD vs sCJD, and spAD vs sCJD). The dashed lines indicate Student's *t*-test cut-off, the data points above the dashed lines represent proteins having a p-value < 0.05 and FC > ±1.5 as significant hits, which are depicted in red (for enriched) and blue (for depleted). **B)** The heatmap of hierarchical clustering of significantly enriched proteins. The log₂-

transformed expression values were normalized by Z-score for each biological replicate. Horizontal axis indicates the differentially enriched proteins, and the vertical axis shows the biological replicates from all the groups. Green denotes depleted proteins, red represents enriched proteins. **C)** The heatmap of hierarchical clustering from SWATH-MS of rpAD and sCJD-MM1 and -VV2 subtypes, showing similarities in RBPome candidates between rpAD and sCJD-MM1 subtype.

3.1.2 Functional categorization of RBP candidates

To gain functional insights into the identified RBP candidates, classification based on Gene Ontology (GO) annotations was performed (Fig. 13B). The analysis was performed in the GO domains, including "Biological Process and Molecular Function". The results of this analysis identified several categories that were differentially enriched in various disease groups.

To compare the functional profiles and to visualize the relative enrichment of different categories in all the groups (Fig. 13B), the protein counts belonging to GO terms from each group were uploaded to Perseus software. The variation in each term across the groups was calculated by a Z-score. The heatmap is representing disease groups across the columns (vertical axis) and functional terms in the rows (x-axis). The enriched terms in comparison to other groups are indicated by red, while depleted terms are indicated by green.

A large number of identified proteins were related to cellular and metabolic processes, RNA metabolism, and immune and stress responses, among other activities. Consistent with the analysis of the RNA-binding proteome, many proteins were annotated as having nucleic acid, nucleotide, RNA- and/or DNA-binding activity as well as transcriptional regulatory and transporter activities (Fig. 13B). A variety of proteins were annotated as having catalytic activity (including lipid and carbohydrate metabolic enzymes and protein modulating enzymes), structural molecule and signal transducer activity (Fig. 13B).

Based on clustering of GO terms, rpAD and sCJD groups were segregated into a single cluster, showing similarities in the enriched functional terms (Fig. 13B). The most enriched processes for both groups included response to stress (e.g. response to oxidative stress and cellular response to stress), metabolic process and catalytic activity, among others (Fig. 13B). The signal transducer activity was particularly enriched in the spAD group when compared with other groups. Additionally, one cluster of cellular processes, including cellular homeostasis, localization and binding, was particularly enriched in the rpAD group, showing more aggressive disturbances in the overall cellular homeostasis of the rpAD group. Taken together, these results suggested that the RBPome aberrations are an integral part of the pathological features of particularly the rapidly progressive forms of dementia.

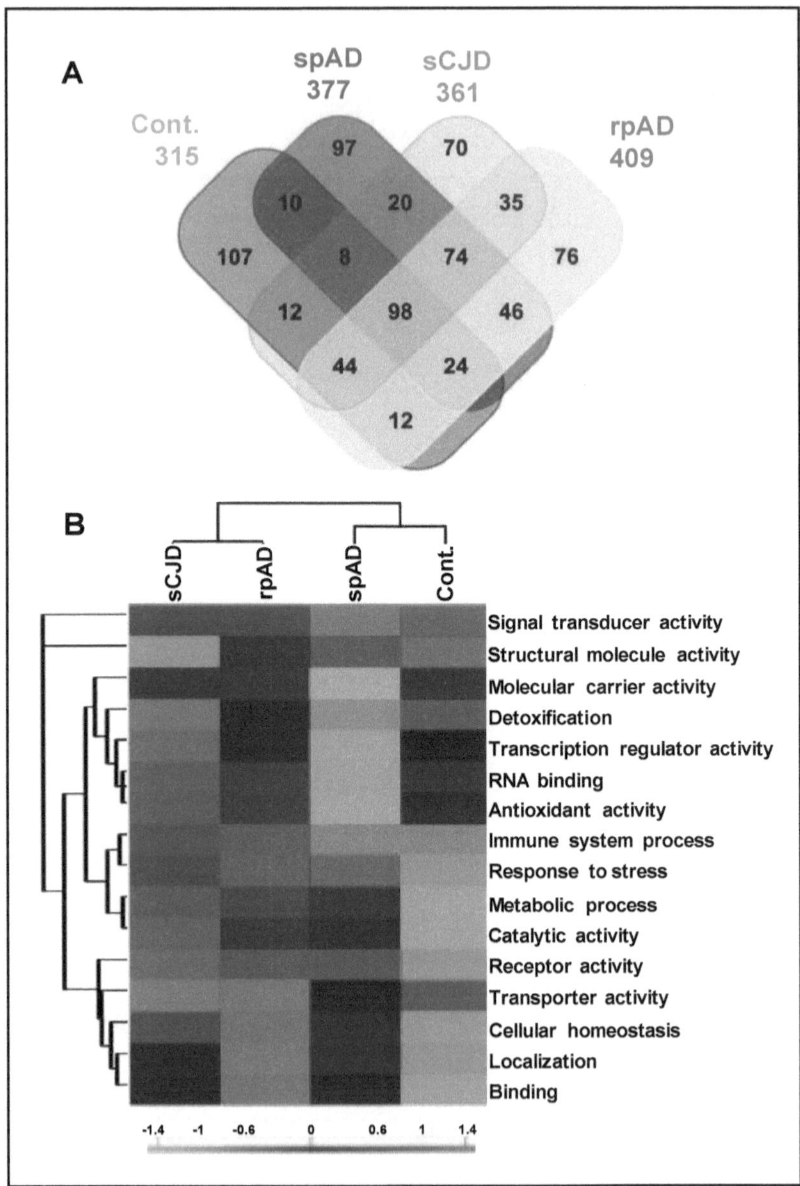

Figure 13: Comparative RNA-binding proteome profiling and functional categorization of MS results. A) Venn diagram showing similar and unique RBP candidates in each group. There were ninety-eight proteins common among all the groups. There were total of 315 proteins in the control, 377 in the spAD, 361 in the sCJD and 409 in the rpAD group. Of these, 107 were only uniquely present in the control, 97 were only found in the spAD, 70 in the sCJD and 76 in the rpAD group. **B)** Functional categorization of MS results: RBP candidates

identified in each group were analysed for associated GO terms. The protein count associated with the GO terms from each group was uploaded to Perseus software, to prepare heatmap showing relative enrichment of different functional categories across all the groups. The variation in each term across the groups was calculated by Z-score. The heatmap is representing disease groups across the columns (vertical axis) and functional terms in the rows (x-axis), with red indicating an enriched category as compared with other groups, and green indicating depleted terms.

3.1.3 Classification of known and putative/novel RBP candidates

Candidates for RNA-binding protein were classified as previously demonstrated (Beckmann *et al.*, 2015). In order to discriminate *bona-fide* RBPs from putative RBP candidates, identified proteins were searched for RNA-binding annotation in the Uni-ProtKB database. Two categories were organized to give clarity to the data, as described previously (Zhang *et al.*, 2016) (Fig. 14A). The category I (*bona-fide* RBP candidates: red bars) contains proteins that were annotated as RNA-binding in the UniProtKB database. There were 64 RBPs in the control group, 64 in spAD, 77 in rpAD and 70 in sCJD that were annotated as RNA-binding proteins (Fig. 14A). The category II (putative/novel RBP candidates: black bars) contains the rest of the proteins that have neither a known RNA-binding domain (RBDs) nor a known linkage to the RNA. The putative RBP candidates include metabolic enzymes and translation factors among others. It should be noted that some of the identified candidate proteins may not have RNA-binding activity themselves, but, instead, interact with other proteins that do, and are hence identified in the RBPome.

3.1.4 Prion-like domain (PLD) prediction

Prion-like domains are often present in RBPs, which carry out protein aggregation in neurodegenerative diseases, e.g. amyotrophic lateral sclerosis (ALS), and, more recently, in AD (King *et al.*, 2012; Wolozin, 2012; Vanderweyde *et al.*, 2016). To evaluate the presence and functional implication of PLD-containing proteins, the identified RNA-binding proteomic candidates were analysed using highly stringent computational algorithms (Batlle *et al.*, 2017; Michelitsch and Weissman, 2000; Toombs *et al.*, 2012; Zambrano *et al.*, 2015). In this study, the web-based PLAAC database was used to identify domains with prion-like amino acid sequences. The input sequences are ranked by several types of summary scores. Criteria for a protein to have a PLD consisted of four requirements: it had to 1) be rich in Q/N-sequences, 2) exhibit disorderliness according to the PAPA algorithm (Prilusky *et al.*, 2005), 3) have compositional similarity with yeast prion domains, and 4) contain short amyloidogenic stretches able to carry out their self-assembly into an amyloidogenic state. Applying these criteria, we identified twenty-four PLD-containing proteins in the current study (Table 10).

Table 10: Prion-like domain prediction score of RBPs from PLAAC database.

Experimental group	SEQid	COREscore	PAPAprop
Cont., sCJD	sp|Q12906|ILF3	30.688	0.092
Cont. rpAD, sCJD	sp|P08247|SYPH	25.009	-0.105
Cont. spAD, sCJD	sp|Q5D862|FILA2	24.633	0.109
Cont., spAD, rpAD, sCJD	sp|P17600|SYN1	20.311	-0.054
Cont, sCJD	sp|O43390|HNRPR	17.059	-0.049
Cont., sCJD	sp|P09012|SNRPA	11.165	-0.146
spAD, rpAD	sp|Q9UPA5|BSN	22.864	-0.044
spAD, rpAD, sCJD	sp|P20073|ANXA7	17.868	-0.033
spAD, sCJD	sp|Q92945|FUBP2	15.926	-0.065
spAD, rpAD, sCJD	sp|P02671|FIBA	12.12	0.033
rpAD, sCJD	sp|Q14103|HNRPD	30.124	0.164
rpAD, sCJD	sp|P23246|SFPQ	28.671	-0.1
rpAD	sp|P04156|PRIO	14.844	0.02
sCJD	sp|Q92734|TFG	38.015	-0.005
sCJD	sp|Q01844|EWS	34.368	0.057
sCJD	sp|P22626|ROA2	30.362	0.043
sCJD	sp|P09651|ROA1	28.381	0.093
sCJD	sp|Q8WUM4|PDC6	26.882	0.049
sCJD	sp|P50995|ANX11	19.803	-0.059
sCJD	sp|Q9UBV8|PEF1	19.787	0.005
sCJD	sp|P17931|LEG3	17.868	-0.033
sCJD	sp|P49840|GSK3A	11.251	-0.103
sCJD	sp|P14678|RSMB	10.958	-0.121
sCJD	sp|O60506|HNRPQ	8.799	-0.019

PLAAC: Prion-like amino acid composition, SEQid: sequence ID from fasta file, COREscore: max sum of PLAAC LLRs, PAPAprop: max score of PAPA prion propensities (Toombs *et al.*, 2012), Cont.: control, spAD: sporadic Alzheimer's disease, rpAD: rapidly progressive Alzheimer's disease, sCJD: sporadic Creutzfeldt-Jakob disease.

Among all the identified PLD-containing proteins, splicing factor proline and glutamine rich (SFPQ) with a PLD score of 28.67 was an interesting target as this protein was specifically enriched in rapid forms of dementia (rpAD and sCJD). The detailed architecture of the SFPQ prion-like domain is described in the figure (Fig. 14B).

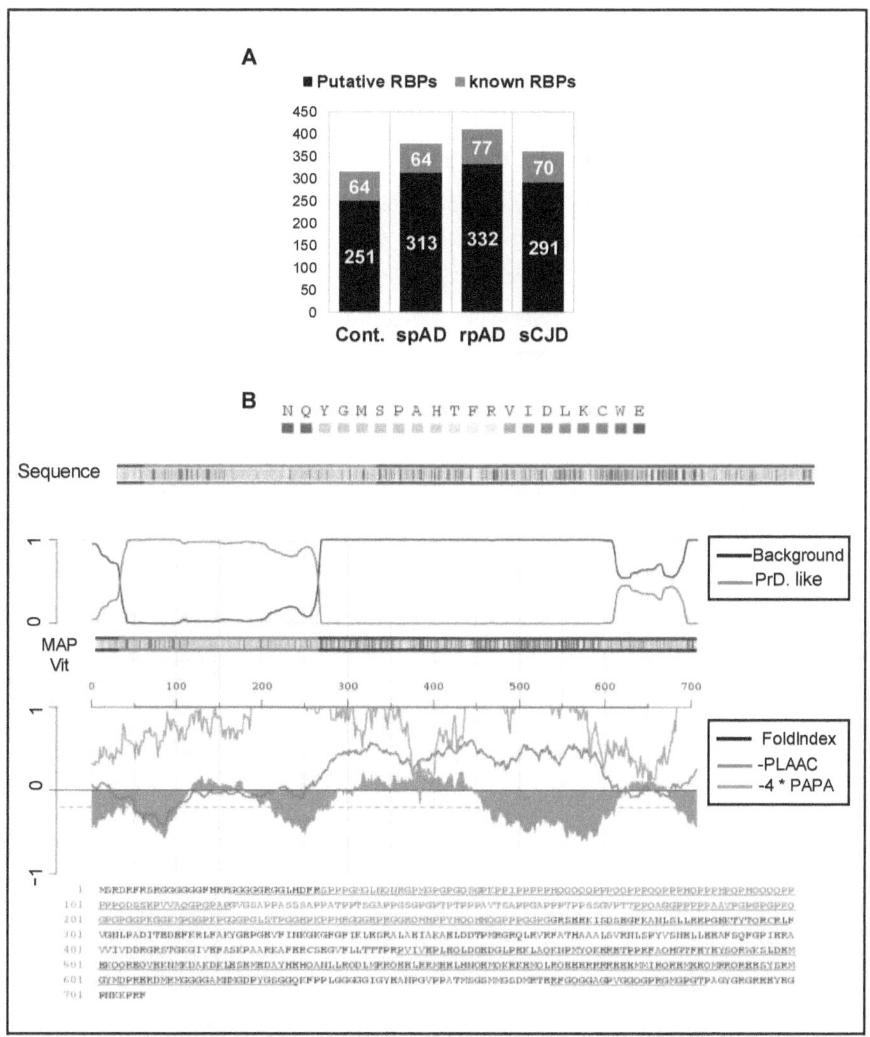

Figure 14: Identification of *bona-fide* and putative RBP candidates. A) The proteins identified in the proteomic study were searched for RNA-binding annotation in the UniProtKB database. The bar graph shows two categories of identified protein candidates. The category I (red bar) indicates *bona-fide* RBPs (known), and category II (black bar) represents potential novel/putative RBP candidates from each group. **B)** Identification of SFPQ-prion-like domain by PLAAC database. The amino acid composition of SFPQ is represented in colour-coded boxes. The red line in the top panel represents the probability of having a prion domain against the background. The plots in the middle panel show fold-index scores in grey (Prilusky *et. al.*, 2005), the log-likelihood (LLR) ratio scores in red (Alberti *et. al.*, 2009), and the predicted prion propensity (PPP) in green (Toombs *et. al.*, 2010). Negative scores represent disorder and prion propensity, dashed green line is indicating the cutoff value of PPP > 0.05. The bottom panel shows the primary sequence of SFPQ with the PLD in red (Alberti *et. al.*, 2009).

3.1.5 Target candidates from proteomic study

From the proteins which had prion-like domains (as described in the Table 10), the SFPQ was selected (approach B). This protein was specifically identified in the rpAD and sCJD groups and exhibited a high score for PLD. This high score of PLD is a crucial factor contributing to the pathophysiology of RBPs. Given the involvement of stress granules in the AD, valosin-containing protein (VCP) was selected from results using the approach A. The valosin-containing protein plays an important role in the formation and clearance of SGs. Valosin-containing protein was also specifically enriched in rapid forms of dementia (rpAD and sCJD).

One interesting finding evident from the proteomic study was the enrichment of the microtubule-associated protein tau (MAPT), a major hallmark of AD in our RBPome dataset, suggesting an intimate link of RBPs to tau-related pathological mechanisms. The interactive association between tau, SFPQ and VCP led us to investigate the role of these target candidates in the disease. Furthermore, TIA-1 (T-cell intracellular antigen 1) was also added to the candidate list for further investigation (Fig. 15), given that its role in tau aberrations in AD (Apicco *et al.*, 2018), although it was not detected in our MS dataset, most likely due to the hydrophobic nature of the protein.

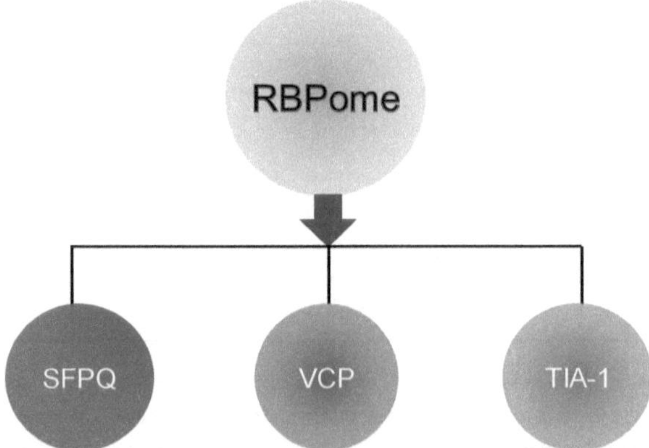

Figure 15: Candidate hits from proteomic investigation. Splicing factor proline and glutamine rich (SFPQ) (approach B) was selected, because the protein was specifically enriched in rpAD and sCJD, and, in addition, bearing high score for PLD; which is very crucial factor contributing to pathophysiology of RNA-binding proteins. Given the recent role of stress granules in AD, valosin-containing protein was selected (approach A). This protein was significantly enriched in abundance only in diseased groups and has a role in the formation and clearance of SGs. Due to involvement of TIA-1 protein in the pathophysiology of tau, it was also added to the target candidate list, although it was not detected in our dataset.

In order to more accurately illuminate the specific mechanisms of how selected targets (particularly SFPQ) interact with other major cellular components (e.g. tau) in the development of AD, and to further ascertain whether the aforementioned selected approaches were appropriate for the target selection, we began to explore these proteins more in-depth. Based on research which has shown that neurons are particularly vulnerable to aberrant dosage as well as the dynamics of RBPs (Conlon and Manley, 2017), the first set of analyses focused on the pathological characterization of these target proteins in the postmortem brains, using various techniques including immunoblotting, immunohistochemistry, and qRT-PCR analysis. Then, these target candidates were translated in the cellular and animal models to find out mechanistic links with pathological features of the disease.

3.2 Pathological characterization of target RBP (SFPQ) in the postmortem human brain

3.2.1 SFPQ is dysregulated in rpAD and sCJD brains

To investigate the differential expression of SFPQ in Alzheimer's and prion diseases, immunoblotting analysis was performed in the frontal cortical brain tissues of human brains from the spAD, rpAD, sCJD-MM1, and sCJD-VV2 subtypes as well as non-demented controls. Information on all the subjects is listed in annexure data tables (Tables 11 and 12). Immunoblotting analysis showed that the expression of SFPQ was reduced in the frontal cortex of both rpAD and sCJD patients compared with controls (Fig. 16A).

For SFPQ expression in spAD patients, a decreasing trend was observed, though that was not significant. The band intensities of SFPQ were normalized to the corresponding levels of GAPDH by calculating the intensity ratio of the GAPDH bands. The densitometric analysis revealed that SFPQ levels were remarkably decreased in both subtypes of sCJD (MM1 and VV2) compared with the control subjects (Fig. 16B).

A significant reduction was observed for TIA-1 levels in spAD and both subtypes of sCJD (MM1 and VV2), when compared with control subjects (Fig. 16A and C). In contrast, the levels of TIA-1 were significantly increased in rpAD patients in comparison to patients in the spAD group. Defects in VCP have been linked with aberrant dynamics of SGs, both as impaired formation and clearance of SGs (Buchan *et al.*, 2013). Next, we sought to examine the levels of VCP in the diseased and control subjects. An increasing trend was observed for VCP levels in rpAD and sCJD-VV2 subtype (Fig. 16D) compared with the respective controls, suggesting an impaired SG dynamic.

Overall, the dysregulated target proteomic signatures suggest a compromised multi-functionality of these proteins, which can start a cascade of aberrant signaling in the neurons and contribute to neurodegeneration.

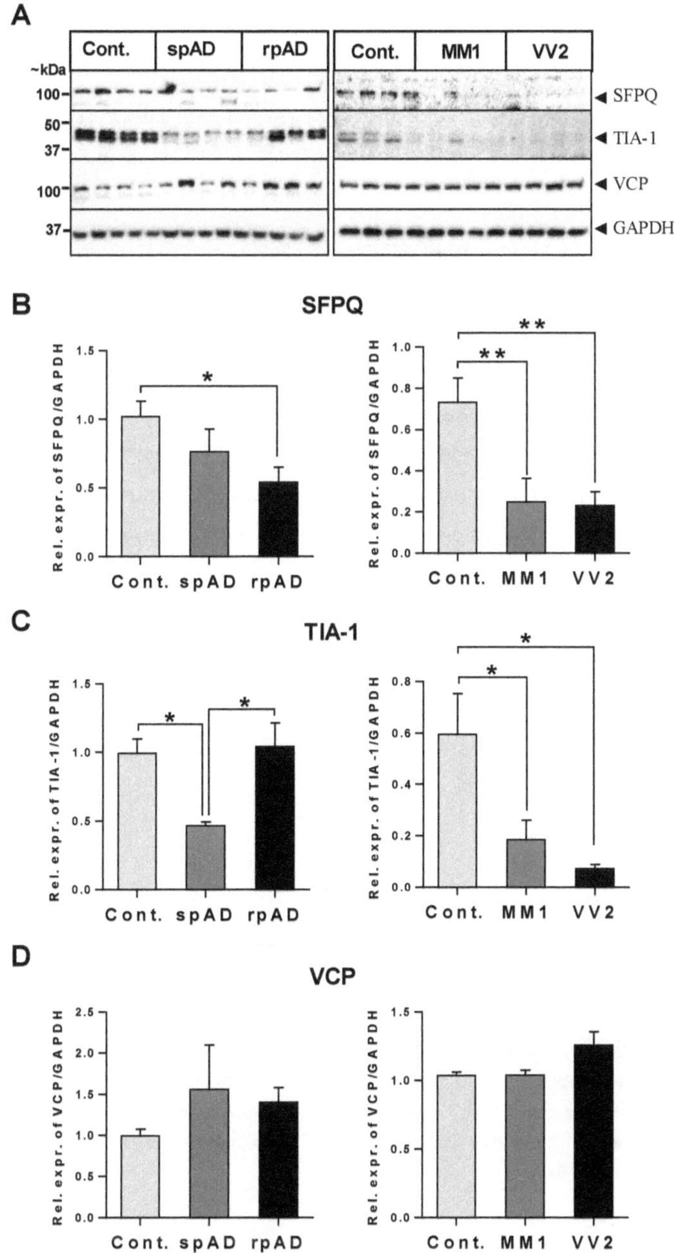

Figure 16: Differential expression analysis of SFPQ, TIA-1, and VCP: A) Immunoblotting analysis was performed with frontal cortical human brain tissues from spAD (n = 7), rpAD (n = 7), sCJD (MM1 and VV2 subtypes,

n = 8) and non-demented controls (n = 8). Representative immunoblot images are shown. **B-D)** The densitometric analysis of SFPQ, TIA-1 and VCP. Protein amounts were normalized to corresponding bands of GAPDH. The levels of SFPQ were significantly reduced in rpAD and the indicated sCJD subtypes. Expression of TIA-1 at the protein level was significantly reduced in the spAD and sCJD groups, while it was increased in rpAD as compared with the spAD group. For VCP, no significant changes were observed. One-way ANOVA was conducted, followed by Tukey post-hoc test for multiple comparisons, *p < 0.05, **p < 0.01.

Dysregulated protein levels of SFPQ and TIA-1 prompted us to investigate the mRNA levels of these target proteins in the diseased brains. Quantitative RT-PCR was used to examine mRNA levels in spAD, rpAD and control subjects. The expression levels were normalized to GAPDH. Contrary to a reduced expression at the protein level in rpAD, mRNA levels of SFPQ were significantly elevated in the rpAD group compared with control and spAD groups (Fig. 17A). For both TIA-1 and VCP mRNA levels, a significant increase was observed in rpAD subjects, in comparison with control and spAD subjects, as was observed for protein expression (Fig. 17B and C).

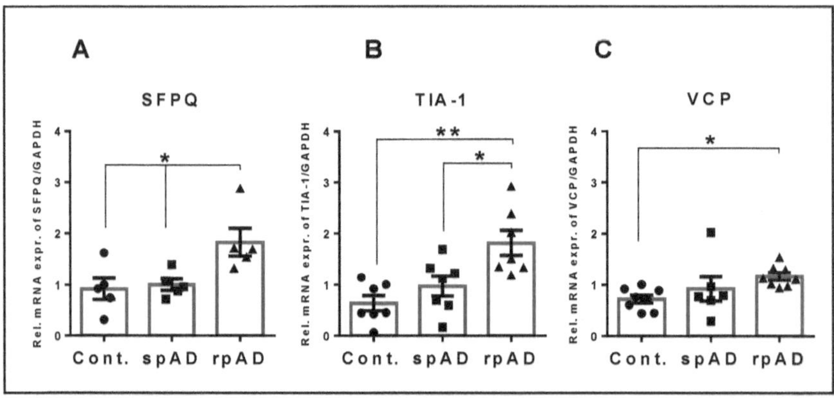

Figure 17: Expression of SFPQ, TIA-1, and VCP at mRNA level. A-C) Expression of SFPQ, TIA-1 and VCP at mRNA level was analysed in spAD, rpAD, and controls using qRT-PCR. The expression levels of mRNA were normalized to GAPDH. The comparative C_t method ($2^{-\Delta\Delta Ct}$) was used for calculation of relative mRNA levels (Livak and Schmittgen, 2001). One-way ANOVA followed by Tukey post-hoc analysis for multiple comparisons, *p < 0.05, **p < 0.01, (n = 5-8).

3.2.2 SFPQ mislocalization and co-localization with SG marker TIA-1 in the rpAD brain

Mislocalization of several RBPs has been described as a pathological feature in many neurodegenerative diseases (Barmada *et al.*, 2010; Bishof *et al.*, 2018; Neumann *et al.*, 2006; Vance *et al.*, 2013). Immunoblotting analysis revealed dysregulation of SFPQ in the human brain from patients diagnosed as rpAD and the two sCJD subtypes. Next, we sought to explore the localization of SFPQ in the human brain. In order to examine the localization of SFPQ, we immuno-stained the brain tissues from spAD, rpAD, and control subjects with antibodies specific for SFPQ. For this analysis, a separate cohort was used and the clinical data of the fifteen cases are listed in the annexure data table (Table 13). In most cells of the control and spAD brains, SFPQ was localized in the nucleus (Fig. 18A) which is the typical localization of SFPQ, as reported previously (Lu *et al.*, 2018; Meissner *et al.*, 2000). Interstingly, in rpAD cases, SFPQ was massively depleted from the nucleus. A ring-shaped SFPQ was observed around the nucleus (Fig. 18A). Dislocation of SFPQ was observed in 91% cells in rpAD, compared with 51% in spAD and 43% in controls (Fig. 18B). These results demonstrate that nuclear depletion and dislocation of SFPQ occurred in both patient and control tissues; however, the dislocation rate was specifically higher in rpAD brains.

The mislocalization of SFPQ, particularly in the rpAD cases, raised an interesting question of whether or not nuclear depletion of SFPQ was associated with its cytoplasmic accumulation. Recently, the mislocalization of some nuclear factors has been linked with their cytoplasmic accumulation in the stress granules (Barmada *et al.*, 2010; Vance *et al.*, 2013). To test this possibility, we double-labelled the tissues from diseased and control subjects with primary antibodies specific for SFPQ (red) and TIA-1 (green): a classical marker of SGs (Fig. 18A). Sudan black was used to quench lipofuscin fluorescence, as differentiation of SG reactivity from lipofuscin can be quite challenging (Liu-Yesucevitz *et al.*, 2010; Vanderweyde *et al.*, 2012). Treatment with Sudan black highlights consolidated cytoplasmic TIA-1 reactivity, which tends to show strong fluorescence in the SGs, compared with nuclear signal which is quenched by Sudan black. Hence, it's difficult to observe nuclear TIA-1 in these Sudan black-treated tissues (Fig. 18A).

TIA-1 reactivity was also observed in control cases with normal cognition, as has been noted previously (Vanderweyde *et al.*, 2012), and a partial co-localization was also observed with SFPQ (Fig. 18A). In case of spAD subjects, a moderate co-localization was observed. For rpAD cases, a complete co-localization was demonstrated for SFPQ and TIA-1 in the perinuclear/cytoplasmic area (Fig. 18A).

To measure quantitative association between SFPQ and TIA-1, co-localization analysis was performed with ImageJ and FIJI (Coloc 2 plugin) software. Multiple methods were used to provide a quantitative measure of the extent of co-localization. We analysed the co-localization of SFPQ and TIA-1 using Pearson's correlation coefficient (rP) and Threshold Mander's coefficient (tM) (Bolte and Cordelieres, 2006), together with intensity correlation analysis (ICA). Intensity correlation analysis (PDM plots) showed highest degree of co-localization between SFPQ and TIA-1 in the rpAD brain, followed by spAD and controls (Fig. 18A). A significant degree of co-localization was observed in rpAD cases, as judged from the two co-localization coefficients rP and tM (Fig. 18C and D). The value of tM1 shows the overlap of TIA-1 channel pixels with SFPQ channel pixels, which is significantly higher in rpAD when compared with either spAD or control subjects. For tM2 (representing the overlap of SFPQ channel pixels with TIA-1 channel pixels), a trend was observed for rpAD cases, as well.

In summary, quantitative analysis of the brain-tissue staining demonstrated that co-localization between SFPQ and TIA-1 was observed in both spAD and rpAD cases, with a stronger degree of co-localization in rpAD subjects.

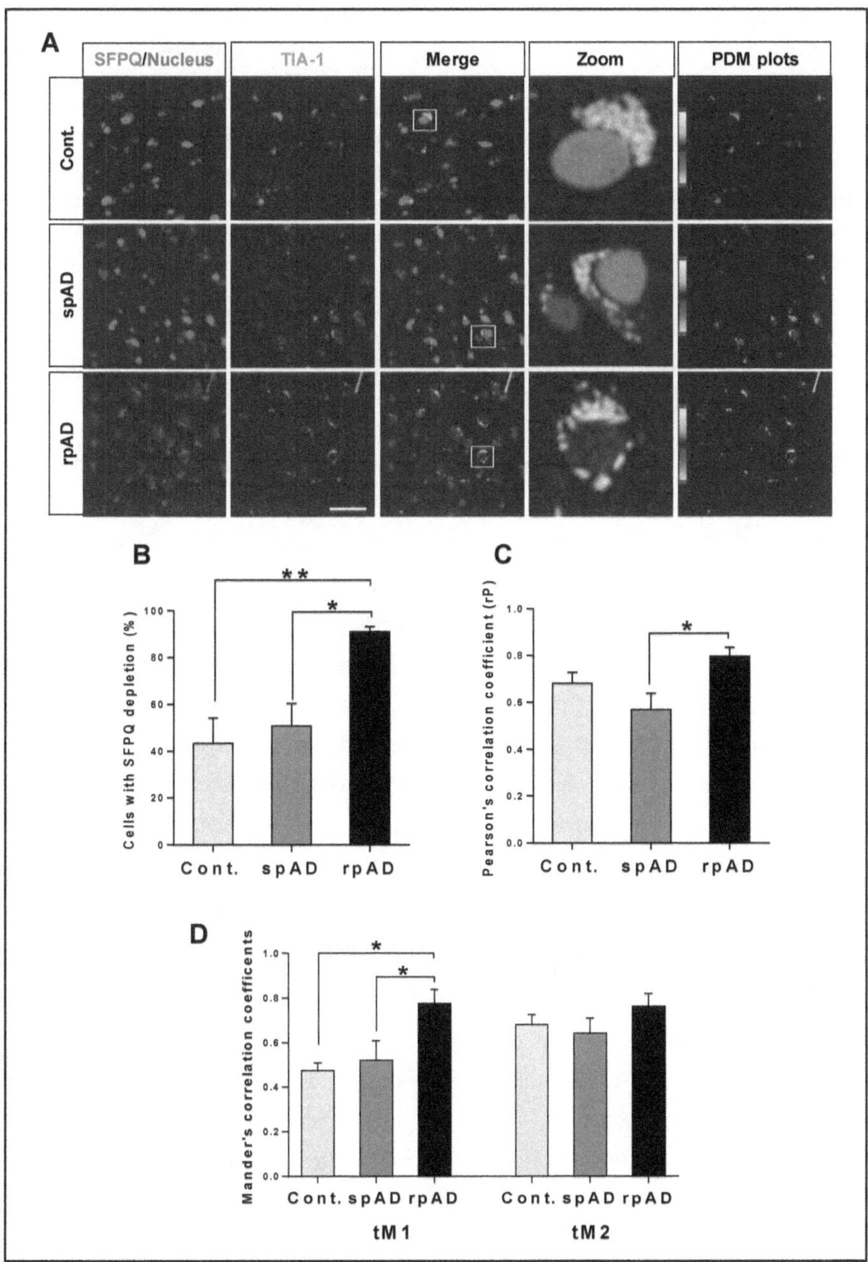

Figure 18: Dislocation/depletion of SFPQ from the nucleus and co-localization with the SG marker TIA-1 in the cytoplasm. A) Co-immunofluorescence of SFPQ (red) and TIA-1 (green) in the human brain of spAD (n = 4), rpAD (n = 4), and controls (n = 5). Cell nuclei were visualized with To-Pro-3 iodide staining (blue). Repre-

sentative images are shown (scale bar = 50 µm). Cells are also shown at higher magnification for a closer look in each group. Intensity correlation analysis (ICA) was performed with ImageJ, showing PDM plots, which are representing highest overlap in rpAD cases, followed by spAD and controls. **B)** Quantification of the cells with SFPQ dislocation (n = 150). **C)** Pearson's correlation coefficient (rP) graph, representing co-localization coefficient between SFPQ and TIA-1. The co-localization between SFPQ (red) and TIA-1 (green) was analysed with FIJI (Coloc 2 plugin) software. **D)** Co-localization analysis with Threshold Mander's correlation coefficients (tM). The value of tM1 shows the overlap of TIA-1 channel pixels with SFPQ channel pixels, and tM2 represents the overlap of SFPQ channel pixels with TIA-1 channel pixels. One-way ANOVA followed by Tukey post-hoc test for multiple comparisons, *p < 0.05, **p < 0.01.

3.2.3 SFPQ is co-localized with phospho-tau in neurofibrillary tangles in the rpAD brain

Cytoplasmic co-aggregation of some splicing factors with tau protein has been reported in both sporadic and familial cases of AD (Bai *et al.,* 2013; Bishof *et al.,* 2018; Diner *et al.,* 2014). To test this possibility, the relationship of SFPQ with tau tangles was explored by immunohistochemical analysis. In frontal cortex tissues from the control subjects, immunoreactivity for SFPQ and phospho-tau (S199) was observed predominantly in the nucleus, with a strong degree of co-localization between both proteins. To further investigate this relationship, typical cells were investigated at high magnification in the indirect immunofluorescence micrographs (Fig. 19A). In the human brain of spAD patients, a strong signal was observed for SFPQ in the nucleus and for phospho-tau in the cytoplasm, with a partial overlap between both in the nuclear and cytoplasmic regions (Fig. 19A). Interestingly, in rpAD subjects, there was a dislocation of SFPQ from the nucleus and immunoreactivity was observed in the cytoplasm. A significant co-localization was evident with phospho-tau tangles in the cytoplasm (Fig. 19A).

In order to measure the association between SFPQ and phospho-tau quantitatively, co-localization analysis was performed with ImageJ and FIJI (Coloc 2 plugin) software. Again, multiple methods were used to investigate the extent of co-localization between SFPQ and phospho-tau. Partial co-localization was observed between SFPQ and phospho-tau in the nuclear region in spAD cases, as was evident from both co-localization methods (rP and tM coefficients). In rpAD cases, two interesting findings were observed. Firstly, co-localization between SFPQ and phospho-tau was significantly increased compared with spAD group (Fig. 19B and C). Secondly, this co-localization was primarily in the cytoplasmic region rather than predominant co-localization in the nucleus, which is typically seen in controls.

In summary, co-immunofluorescence analysis of the affected brain tissues revealed extranuclear distribution of SFPQ and phospho-tau, and co-localization in the cytoplasm, contrary to the predominantly nuclear-based co-localization observed in control cases.

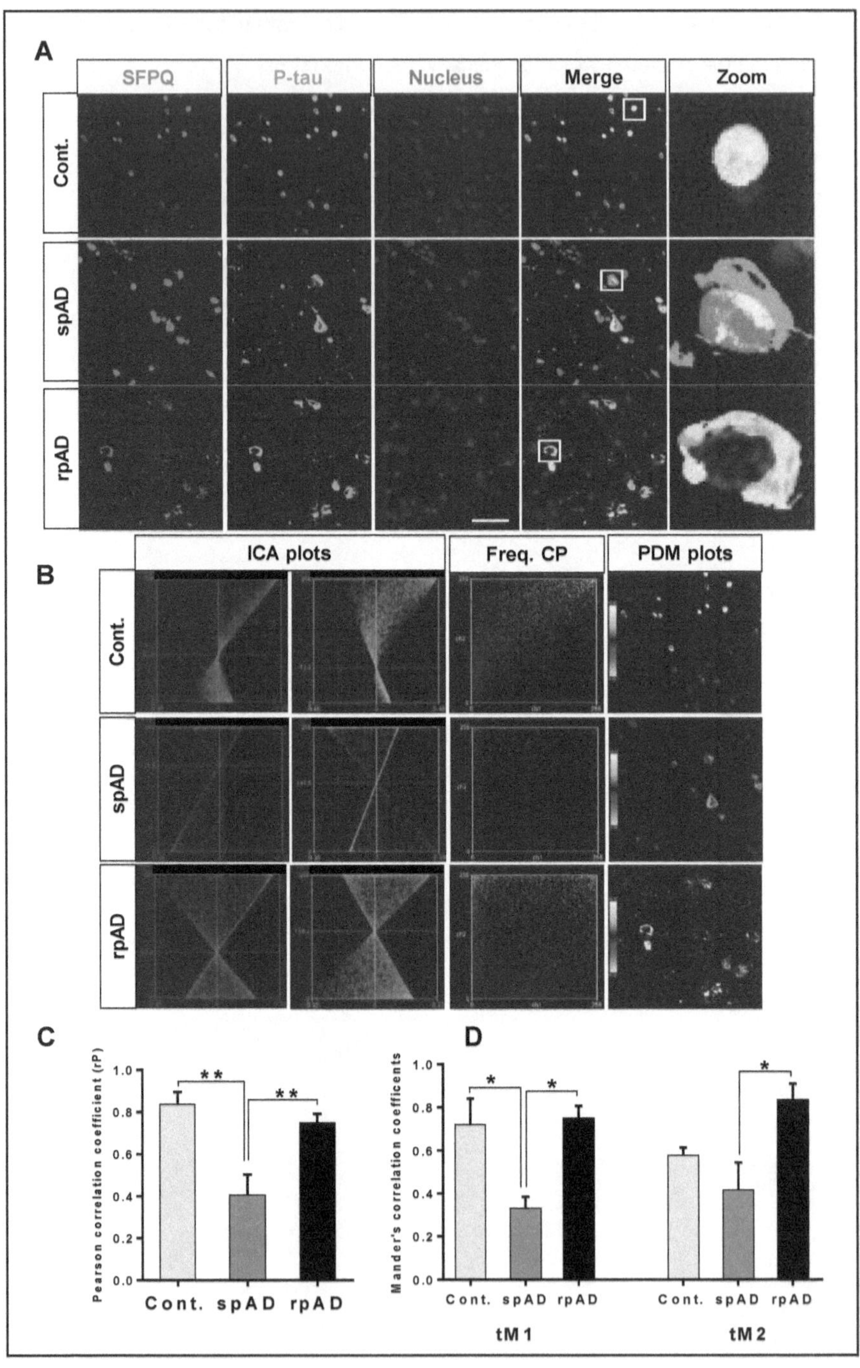

Figure 19: Co-localization of SFPQ with phospho-tau tangles. **A)** Representative images stained with α-phospho-tau (S199) (green) and SFPQ (red) antibodies (scale bar = 50 μm), counter-stained with To-Pro-3 iodide to visualize the nuclei (blue). In control subjects (n = 3), a strong degree of co-localization between SFPQ and phospho-tau was observed in the nucleus at high magnification. In spAD tissues (n = 3), partial co-localization between SFPQ and phospho-tau was observed in the nuclear region. In rpAD cases (n = 3), a strong degree of co-localization was observed between SFPQ and phospho-tau in the cytoplasm, contrary to the nuclear co-localization which was seen in the control subjects. **B)** Intensity correlation analysis (ICA) was performed with Image-J (WCIF plug-in), ICA plots, frequency co-localization plots and PDM plots are displayed for control, spAD and rpAD. **C)** Pearson's correlation coefficient (rP) showing significant co-localization between SFPQ and phospho-tau in rpAD and control cases, as compared with spAD cases. **D)** Threshold Mander's correlation coefficients (tM) representing similar significant co-localization between SFPQ and phospho-tau in rpAD, in comparison to spAD cases. Graphs were prepared with GraphPad Prism (6.01) using One-way ANOVA followed by the Tukey post-hoc test for multiple comparisons, *p < 0.05, **p < 0.01.

The formation of pathological tau tangles is associated with hyperphosphorylation of the tau protein in the AD brain. Expression of total tau and phosphorylated tau (S199) was assessed by immunoblotting analysis (Fig. 20A). Biochemically, no significant changes were observed in the levels of total tau (tau-5), and phospho-tau to tau ratio (Fig. 20B and D). The levels of phosphorylated tau (S199) were significantly increased in the high molecular weight (HMW = 65–250 kDa) range in spAD cases, compared with control subjects (Fig. 20C). A trend was also observed for rpAD, but this relationship was not significant (Fig. 20C).

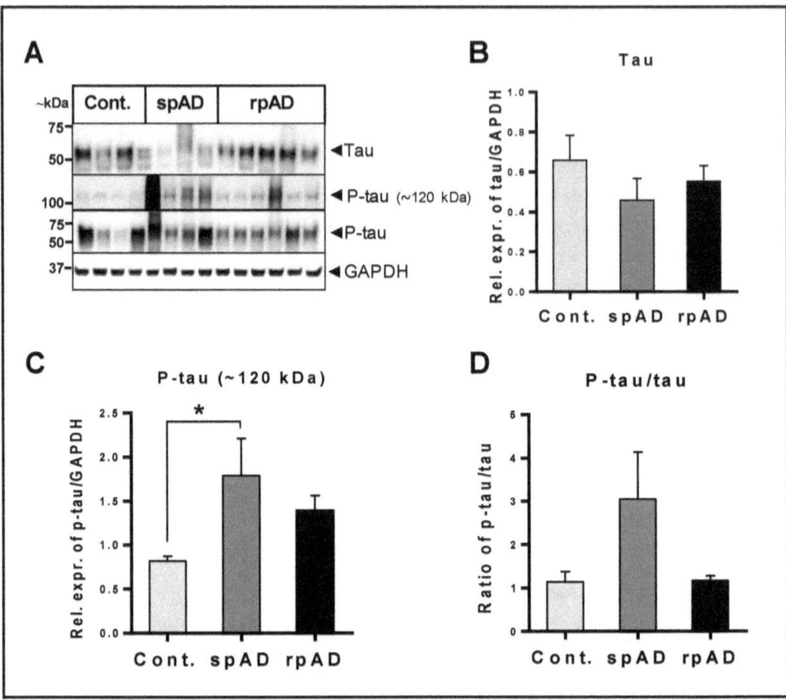

Figure 20: Differential expression analysis of tau and phospho-tau by immunoblotting. A) Representative immuno-blot images for total tau and phospho-tau expression from control, spAD, and rpAD frontal cortical tissue extracts. **(B-D)** Quantification of immunoblotting images for total tau and phospho-tau [low molecular weight-tau (LMW < 65 kDa, and high molecular weight (HMW = 65-250 kDa)]. GAPDH was used as a loading control. There were no significant changes observed for total tau and the ratio of phospho-tau to tau. High molecular weight-range band (~120kDa) of phospho-tau was significantly increased in spAD in comparison to control, while a trend was observed for other bands. Graphs were plotted by GraphPad prism (version 6.01). One-way ANOVA followed by Tukey post-hoc test for multiple comparisons were conducted, *p < 0.05.

3.2.4 Tau oligomers are co-localized with SFPQ in the rpAD brain

Although cytoplasmic tau tangles are a burden for the cell, it has recently been shown that toxic, soluble oligomeric species of tau are the real culprits associated with cognitive decline, neuronal dysfunction, and death (Guerrero-Muñoz *et al.,* 2015; Shafiei *et al.,* 2017). In order to examine the association of SFPQ with oligomeric tau, we co-immuno-stained the cortical sections from control and rpAD subjects, given that there is significant association of SFPQ with tau tangles in the rpAD brain, using SFPQ and anti-oligomeric antibody for tau (T22). Almost no reactivity was observed for oligomeric tau in control cases (Fig. 21A). Interestingly, the co-immunofluorescence analysis revealed a change in the fluorescence pattern of SFPQ, with a high degree of association with oligomeric tau in the cytoplasmic region, in rpAD subjects (Fig. 21A).

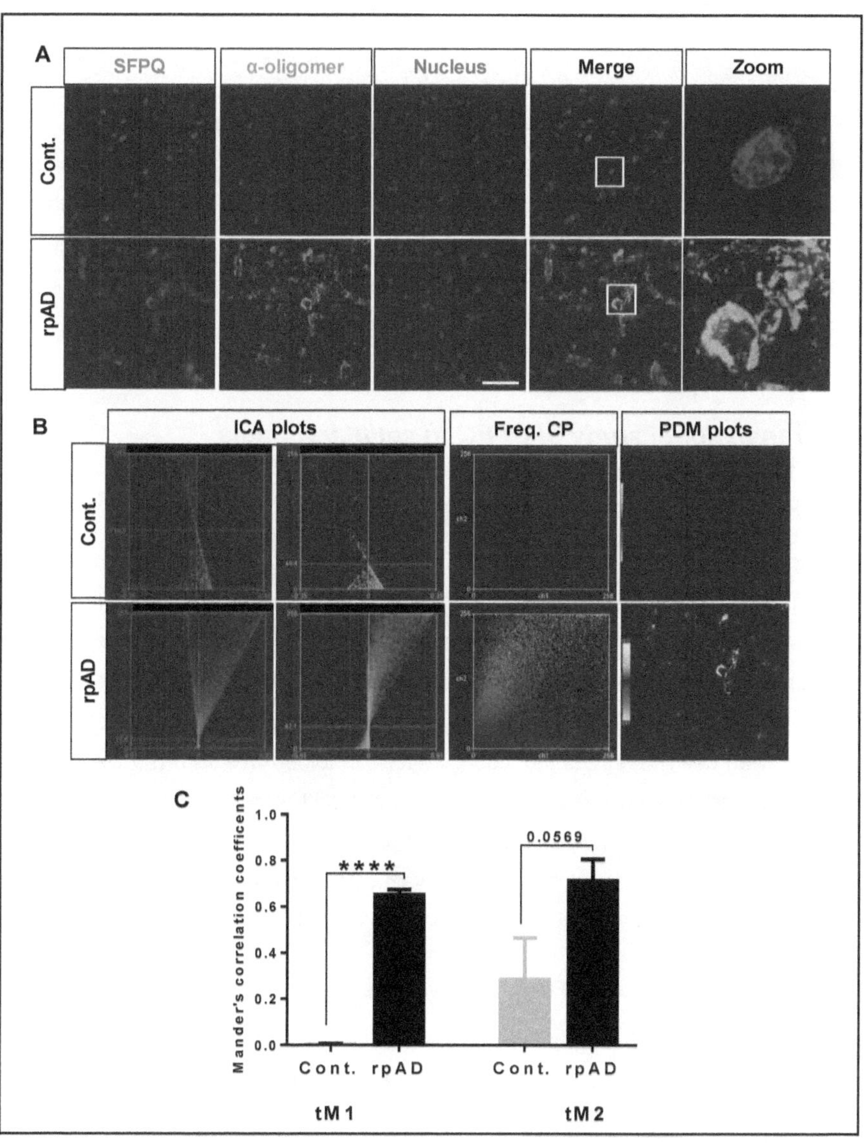

Figure 21: Co-localization of SFPQ with tau oligomers in the rpAD brain. A) Co-immunofluorescence images of control (n = 3) and rpAD (n = 3) cortical sections stained with α-SFPQ (red) and α-Tau oligomeric antibody T22 (green). To-Pro-3 iodide was used for staining nuclei (blue), scale bar = 50 μm. There was no co-localization observed in control cases, while a significant co-localization was observed in rpAD subjects (yellow colour at higher magnification). **B)** Intensity correlation analysis showing ICA plots, frequency co-localization plots and PDM plots for both control and rpAD. **C)** Threshold Mander's correlation coefficients (tM1, tM2) showing significant association between SFPQ and oligomeric tau. Statistical significance was calculated by *t*-test, ****p < 0.0001.

The degree of co-localization was estimated with ICA and Threshold Mander's correlation coefficients (tM1, tM2) (Fig. 21B and C). Significant co-localization was identified by both methods. Splicing factor proline and glutamine rich has a high PLD score and showed co-localization with tau oligomers. This potential interaction between SFPQ and tau oligomers might act as nidus for tau oligomerization and subsequent aggregation.

3.3 Translational study of SFPQ in cellular models

3.3.1 Role of SFPQ towards stress axis

Redistribution of SFPQ from the nucleus into the cytoplasm was observed in the staining of the human brain tissues particularly in rpAD subjects and a co-localization with TIA-1 in the cytoplasm. In order to investigate both the significance of this redistribution and the association with the stress granule protein TIA-1, a cellular model of stress was established in HeLa cells. A well-known oxidative stress inducer, sodium arsenite, was used for stress induction. Several studies have shown a clear link between arsenite-induced oxidative stress and AD pathology in both animal models and a variety of cellular models (McEwen *et al.*, 2005; Resende *et al.*, 2008). Stress granules were analysed, where the viability of both control (untreated) and arsenite-treated cells under stress was not compromised (Fig. 40 in annexure data).

3.3.1.1 Characterization of TIA-1-positive SGs

For visualization of SGs, cells were stained after stress induction with the classical marker and core-nucleating factor of SGs, TIA-1. TIA-1 immunoreactivity was identified in both the cytoplasm and the nucleus, with predominance in the nucleus in the control (untreated) cells (Fig. 22A). Sodium arsenite treatment resulted in the formation of clearly defined cytoplasmic foci which positively stained TIA-1 in more than 80% of cells (Fig. 22A and D). To assess the expression of TIA-1 after stress induction, immunoblotting analysis was performed in both untreated (cont.) and arsenite-treated (stress) cells. A significant increase was observed in the intensity levels of TIA-1 after arsenite-induced cellular stress (Fig. 22B and C).

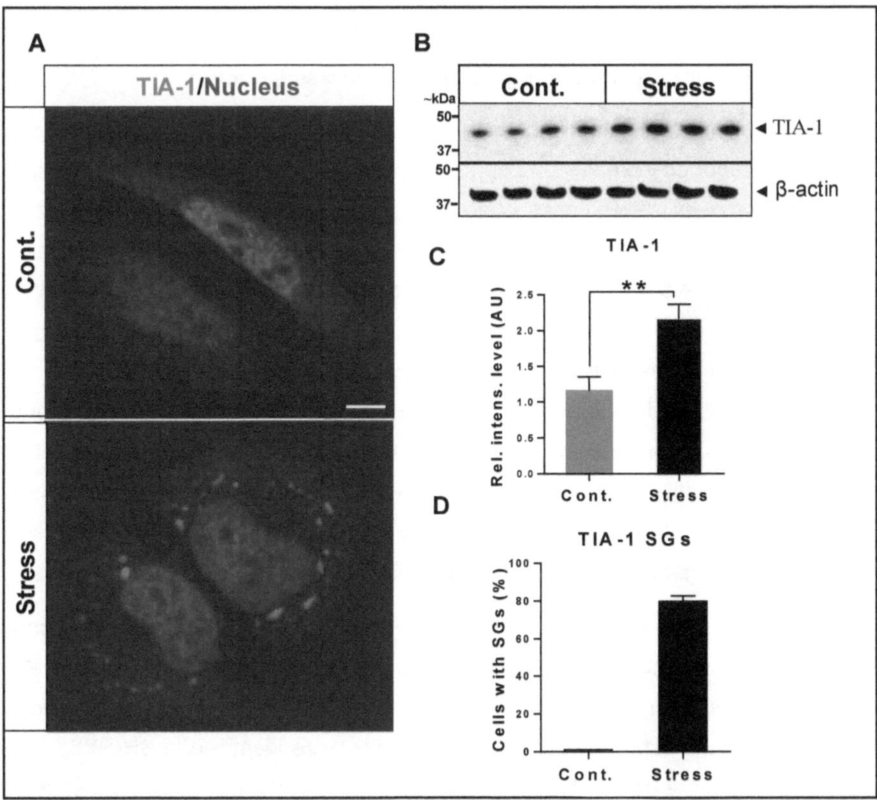

Figure 22: Sodium arsenite induces the formation of SGs in HeLa cells. A) HeLa cells were seeded on glass cover slips in 24-well plates (5×10^4 cells/well) for 24 hrs in DMEM supplemented with 10% FBS and 1% PS. Cells were treated with sodium arsenite (0.6 mM) at 37°C for 1 hr followed by fixation with 4% PFA for 20 min at RT. Using primary antibody specific for TIA-1, stress granules were visualized by staining the classical marker of SGs TIA-1, followed by incubation with secondary antibody AlexaFluor 546. The cells were counter-stained for visualization of nuclei and mounted with immuo-mount mounting medium, scale bar = 10 um. **B and C)** Levels of TIA-1 were determined by immunoblotting followed by densitometric analysis. Significance was estimated by t-test, **p < 0.01. **D)** The cells positive for SGs were calculated with FIJI software. More than 80% cells were identified positive for SGs after arsenite treatment.

A significant proportion of our knowledge on these cytoplasmic foci originates from studies in HeLa cells, including 154 reports published between 1999-2014, according to Aulas and Vande Velde (2015). Based on observations from the current study and a literature survey, HeLa cells were found to be the best representative of SGs. Hence, the HeLa cell line was preferred to further investigate the role of the target protein candidates in SG biology.

3.3.1.2 Tau phosphorylation is increased after stress induction

Numerous rodent studies focusing on the role of acute physiological or psychological stress have reported increased phosphorylation of tau after stress exposure (Feng *et al.*, 2005; Korneyev, 1998; Lopes *et al.*, 2016; Papasozomenos, 1996; Planel *et al.*, 2001, Planel *et al.*, 2004; Yanagisawa *et al.*, 1999). Furthermore, several studies in cellular models have also demonstrated that oxidative stress leads to increased tau phosphorylation in neuronal cultures (Su *et al.*, 2010; Zhu *et al.*, 2005). Firstly, we investigated the levels of total tau after stress induction. There was no significant change observed for total tau levels between the control and stress-induced groups (Fig. 23A and C). For phospho-tau (S199), stress treatment induced significant increases in the levels of phosphorylation when compared with untreated control cells (Fig. 23A and D). This was found to be within the HMW range (65–250 kDa), suggesting tau aggregation under stressful conditions (Su, 2010).

To rule out the possibility that this observed increase in the tau phosphorylation in HeLa cells was cell-specific, the status of the tau phosphorylation was also investigated in the neuronal cell line SH-SY5Y after stress induction. Similarly, significant increases in the phosphorylated tau (S199) levels were identified in SH-SY5Y cells (Fig. 41 in annexure data). Overall, these results indicate that oxidative stress treatment increases tau phosphorylation in both cell lines tested.

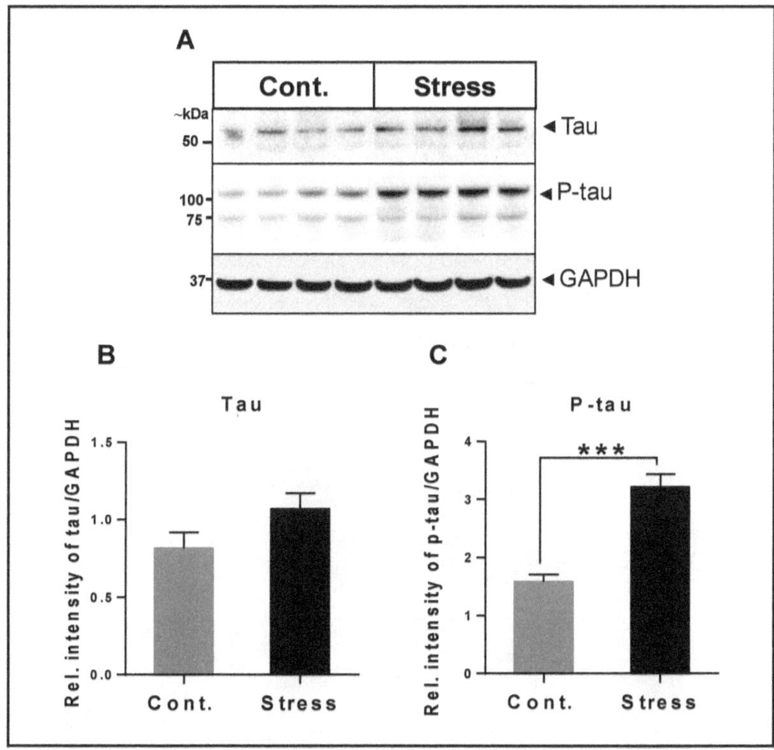

Figure 23: Stress-induced increase in tau phosphorylation. A) Representative immunoblots for total tau and phospho-tau in control (untreated) and stress (arsenite-treated) cells. Cells were plated in 6-well plates ($2x10^5$) for 24 hrs and lysed in cell-lysis buffer supplemented with protease and phosphatase inhibitors. The expression of total tau and phospho-tau was analysed by immunoblotting. Intensity levels were normalized to GAPDH. **B and C)** The densitometric analysis was performed using Image Lab software. Statistical tests were applied in GraphPad prism (version 6.01) with significance ***$p < 0.001$.

3.3.1.3 Tau and phospho-tau are recruited into SGs

Previously, it has been reported that tau co-localizes with the SG markerTIA-1 *in vivo* (Vanderweyde *et al.*, 2012). Next, we studied the relationship between tau and SGs in our cellular model of stress. We examined the association of phospho-tau and tau with TIA-1-positive SGs using immunocytochemistry. Labelling with total tau and phospho-tau antibodies revealed co-localization with TIA-1 cytoplasmic SGs (Fig. 24A and B, high magnification images). Untreated control cells showed a positive signal for tau-5 predominantly in the cytoplasm. For phospho-tau (S199), a positive signal was detected in the cytosol and nucleus, with predominance in the nucleus (Fig. 42 in annexure data), which was in-line with previous studies conducted in HeLa cells (Ibáñez-Salazar *et al.,* 2017; Sjöberg *et al.,* 2006). Such a signal was increased upon oxidative stress treatment. Overall, these findings determine that tau and phospho-tau are recruited into stress granules after stress exposure. The average number of SGs in each cell was estimated with FIJI software. We identified a higher number of phospho-tau-positive SGs compared with tau-positive granules (Fig. 24C). In addition to these observations, it should be noted that the size of the phospho-tau-positive granules was bigger in comparison to tau-positive SGs.

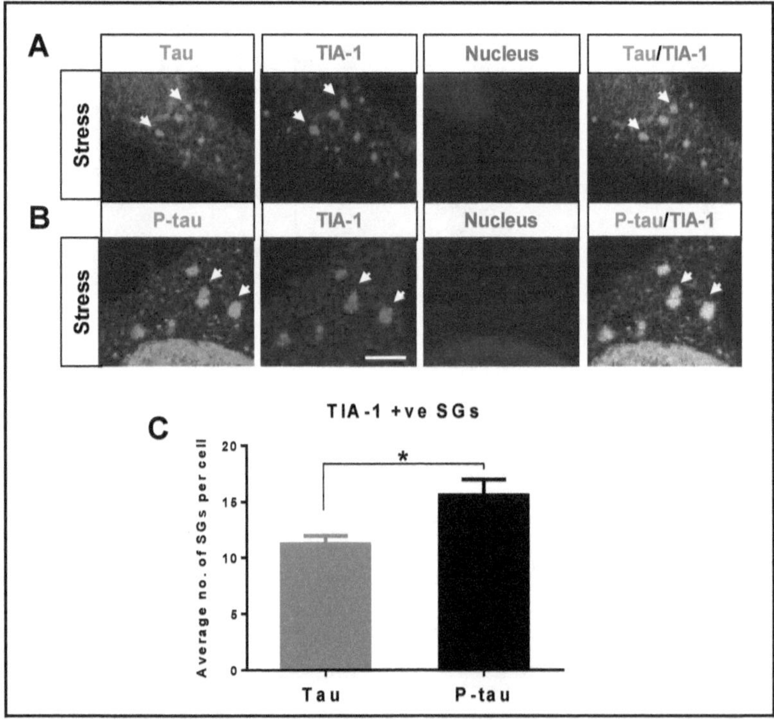

Figure 24: Tau and phospho-tau are recruited into SGs. A and B) Stress was induced with sodium arsenite as above and cells were co-immunoassayed with primary antibodies specific for total tau, phospho-tau and TIA-1, followed by incubation with AlexaFlour 488 and AlexaFlour 546 secondary antibodies. High magnification images showing the expression of tau/TIA-1 and phospho-tau/TIA-1 (Images with lower magnification are shown in annexure Fig. 42). Examples of SGs are indicated by the arrows. **C)** Average number of SGs per cell positive for TIA-1 was estimated both for tau and phospho-tau-positive stress granules using FIJI software. Significance was calculated by *t*-test, *p < 0.05.

3.3.1.4 Endogenous SFPQ redistributes into the cytoplasm and assembles with SGs upon oxidative stress treatment

To ascertain whether endogenous SFPQ forms SGs (as co-localization between SFPQ and TIA-1 was observed in the human brain), we studied the subcellular localization of SFPQ after stress induction. The presense of SFPQ in arsenite-induced cytoplasmic foci was observed by fluorescence microscopy. The protein SFPQ was mainly localized in the nucleus in the control (untreated) cells, which is the typical localization of SFPQ (Fig. 25A). The nuclear localization of SFPQ was not changed after arsenite treatment, but stress-induced redistribution/translocation of SFPQ following arsenite exposure into the cytoplasm and formation of granules in HeLa cells was observed (arrows, Fig. 25A).

To identify whether or not the SFPQ granules which could be observed in the cytoplasm were actually SGs, HeLa cells were co-stained with the SG marker TIA-1. Co-localization was detected between the SG marker and SFPQ (yellow foci in the cytoplasm) (Fig. 25B). The amount of cytoplasmic SFPQ signal was low because labelling only detected endogenous SFPQ, and SFPQ that was present in the cytoplasm was largely in the inclusions. Quantification of SFPQ-positive granules by FIJI software indicated ~20–25 SFPQ-positive stress granules per cell (Fig. 25C). Overall, these results revealed that stress induces cytoplasmic redistribution and formation of SFPQ inclusions, which co-localize with TIA-1 positive SGs.

Furthermore, we examined SFPQ levels before and after stress induction using immunoblotting analysis. A significant increase was identified in the SFPQ intensity levels after stress induction (Fig. 25D and E). Given the role of VCP in stress-granule dismantling and clearance, we next examined the levels of VCP after stress induction. A significant increase in VCP intensity levels was identified in stress-induced cells as compared with control cells (Fig. 25D and F).

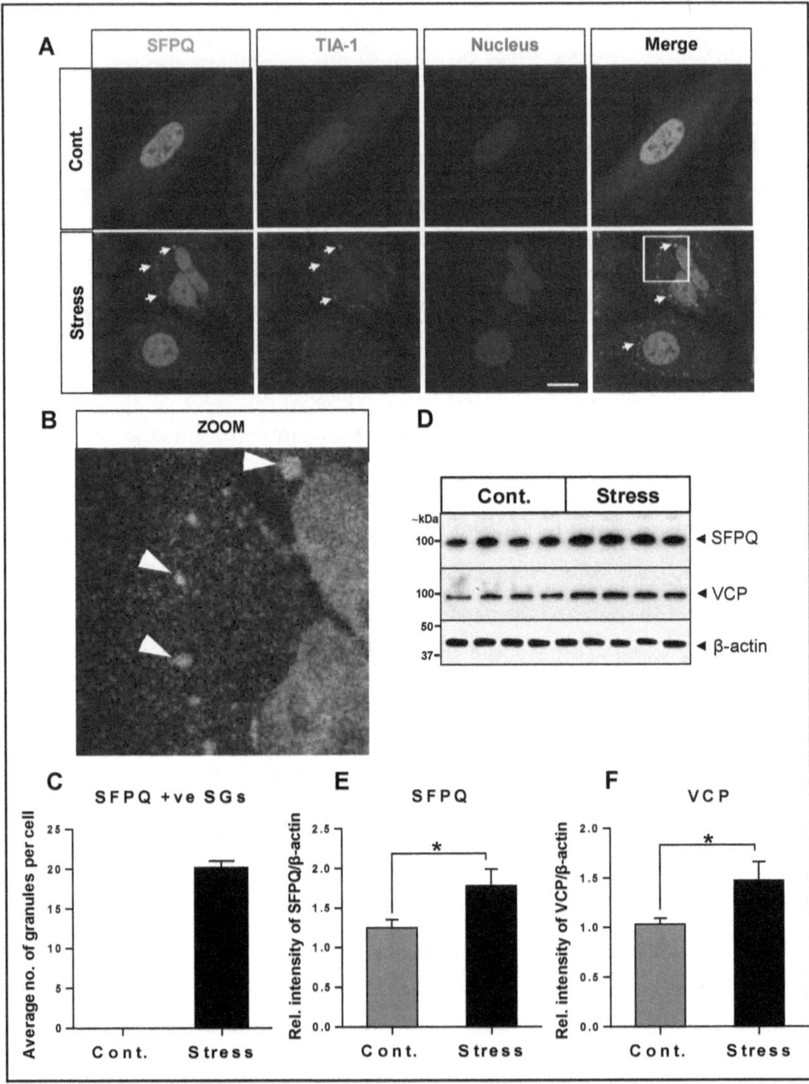

Figure 25: Recruitment of SFPQ into SGs after oxidative stress induction in HeLa cells. A) Localization of SFPQ (green) and TIA-1 (red) was visualized in sodium-arsenite-treated (0.6 mM; 60 min) (stress) and untreated (control) HeLa cells using immunofluorescence microscopy. Cells were counter-stained to visualize nuclei, scale bar = 10 µm. B) Higher magnification image of "A" showing the overlap between SFPQ/TIA-1 in the cytoplasm. C) Average number of SFPQ-positive SGs per cell was calculated with FIJI software. D) Representative immunoblot images after stress treatment. Cells were seeded in 6-well plates (2x10^5) for 24 hrs. Cells lysis was performed with cell-lysis buffer supplemented with protease and phosphatase inhibitors, and expression of SFPQ

and VCP was analysed by immunoblotting. Intensity levels were normalized to β-actin. **E and F)** SFPQ and VCP densitometric analysis. Statistical tests (*t*-test) were applied in the GraphPad prism (6.01), *p < 0.05.

According to a recently described model (Molliex *et al.*, 2015), SG formation is dependent upon liquid-liquid phase separation (LLPS), a property of the RNA-binding proteins containing prion-like domains. We analysed SFPQ protein with catGRAN-ULES algorithm, which predicts the propensity of a given protein to undergo phase separation, and the probability of forming granules. A score of 1.66 was identified for SFPQ showing high probability of SFPQ to form granules. We were also curious to know LLPS properties for TIA-1. Between the two proteins tested, SFPQ (1.66) exhibited a higher score for LLPS than TIA-1 (0.973) (Fig. 26A and B).

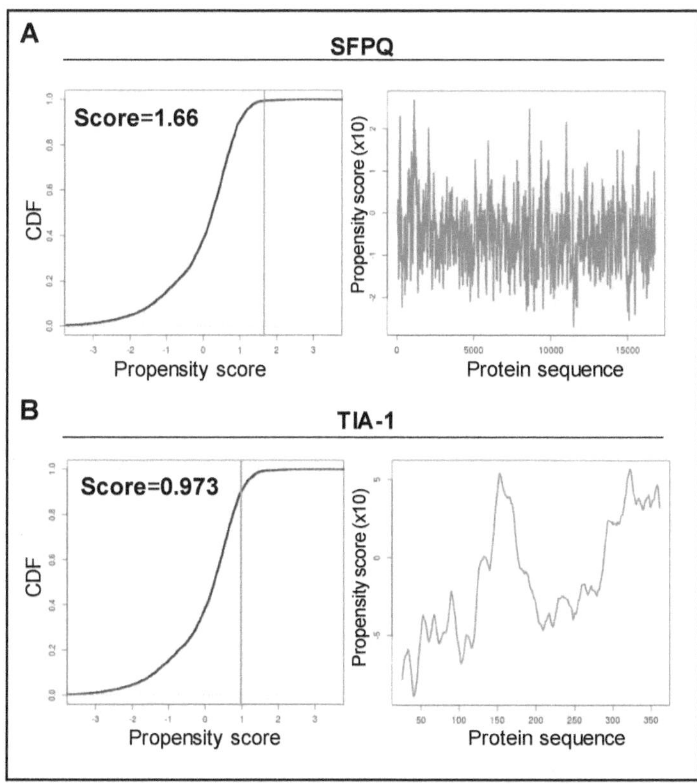

Figure 26: **Liquid-liquid phase separation properties of SFPQ and TIA-1 assessed by catGRANULES algorithm**. SFPQ exhibited a score of 1.66, followed by TIA-1: 0.973, showing higher probability for oligomerization and granule formation.

3.3.1.5 SFPQ co-localizes with tau and phospho-tau in cytoplasmic granules

Recruitment of SFPQ into SGs and localization of tau and phospho-tau in these SGs, raised the possibility that SFPQ and tau might co-localize in SGs and this functional interaction between SFPQ and tau might have implications for neurodegeneration. In the control cells, SFPQ was observed predominantly in the nucleus, while a redistribution was identified in the cytoplasmic-granules in treated (stress) cells. The immunoreactivity of SFPQ was increased after arsenite treatment (stress).

Labelling with tau-5 revealed majorly cytoplasmic staining in the control untreated cells (Fig. 27A). A predominant nuclear signal was detected for phospho-tau. Co-localization of SFPQ with both tau-5 and phospho-tau was observed under basal conditions and in granules upon stress treatment (Fig. 27A and B).

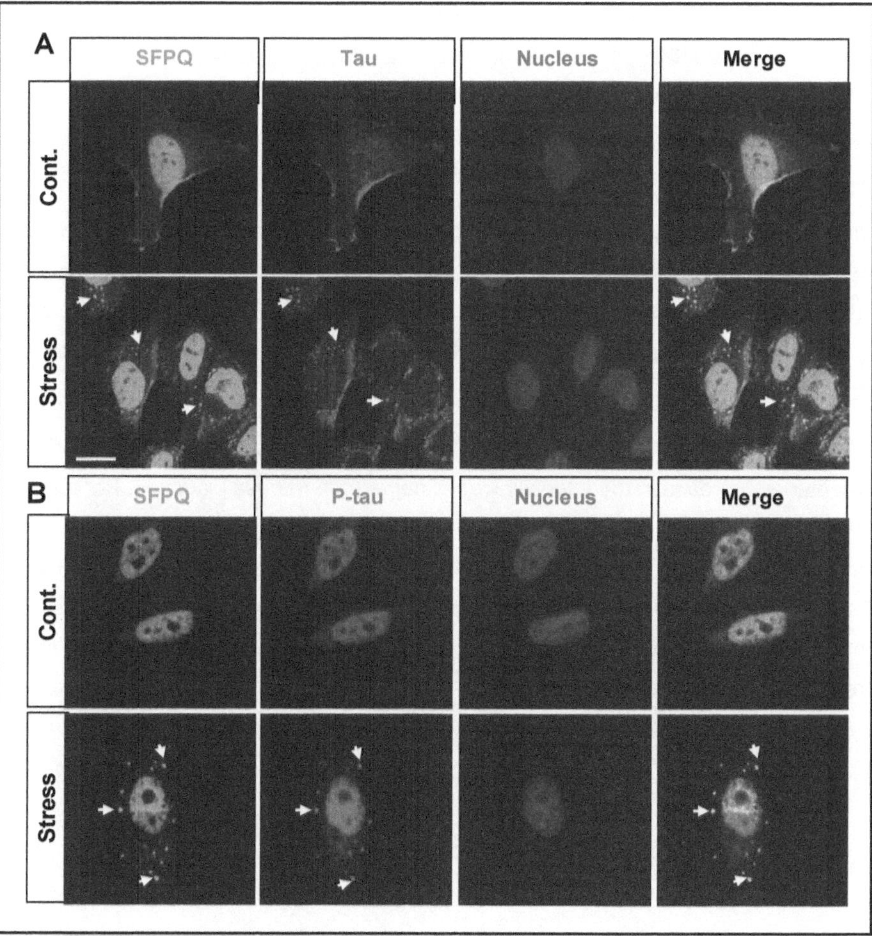

Figure 27: SFPQ co-localizes with tau and phospho-tau in SGs after sodium-arsenite-induced oxidative stress in HeLa cells. A and B) Localization of SFPQ (green), tau-5 and phospho-tau (red) was visualized in control (untreated) and stress (sodium-arsenite-treated) cells by immunofluorescence. Cells were counter-stained to visualize nuclei. Scale bar = 10 µm. Merged micrographs of cells (arsenite-treated) showing the overlap between SFPQ/tau and SFPQ/p-tau in the cytoplasmic granules.

3.3.1.6 Biochemical characterization of stress granule components

Increased intensity levels of SG-associated proteins after stress induction, raised the question of whether or not this was really an increase in their concentration or simply an increase in the intensity levels due to the consolidation of these proteins in SGs. To address this question, we performed subcellular fractionation of control and stress-induced HeLa cells using a Rapid, Efficient and Practical (REAP) method (Suzuki *et al.,* 2010). This method is a very rapid (2 min), non-ionic detergent (NP-40)-based purification method allowing very rapid fractionation of nuclear and cytoplasmic fraction, which is necessary given that nucleocytoplasmic transport of proteins is a rapid process in response to stress.

Immunoblotting analysis revealed an efficient subcellular fractionation with no cross contamination between nuclear and cytoplasmic fractions. The nuclear marker (BRD4) was not detected in the cytoplasmic fractions (Fig. 43 in annexure data). Conversely, the cytoplasmic marker (GAPDH) was not detected in the nuclear fraction (Fig. 43 in annexure data).

After an efficient fractionation, immunoblotting analysis was performed to investigate whether an increase in the intensity was observed after stress induction. For total tau levels, no significant changes were observed in line with the observations from total cell lysates analysis (Fig. 28A and B). Indeed, the REAP method showed a significant increase in the levels of phospho-tau in cytoplasmic and nuclear fractions, also in the HMW range (65-250 kDa: ~120 and 75 kDa), as was observed after stress treatment (Fig. 28A, C and D).

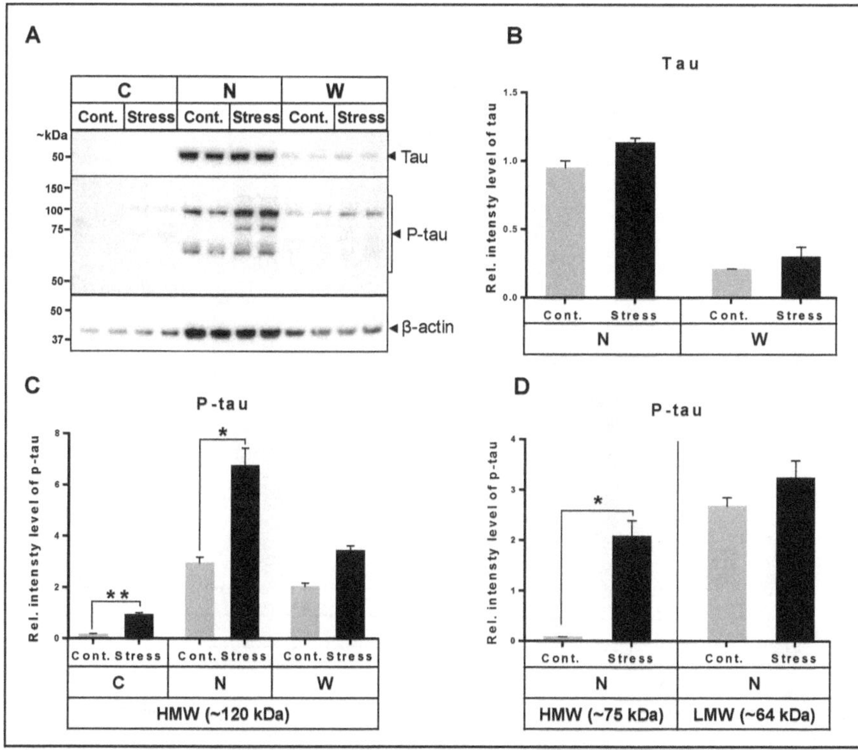

Figure 28: Increased tau phosphorylation and cytoplasmic accumulation after stress treatment. A) HeLa cells were treated with sodium arsenite (0.6 mM, 60 min). Lysates from control (untreated) and stress (arsenite-treated) cells were separated into nuclear and cytoplasmic fractions by REAP method, which were analyzed by immunoblotting with total tau (tau-5) and phospho-tau (S199) antibodies. Intensity values were normalized to β-actin. Isolated fractions were abbreviated as **C**: cytoplasmic extract, **N**: nuclear extract, and **W**: whole cell extract. **B-D)** Quantification of all above proteins showing significantly increased intensity levels for phospho-tau as calculated by *t*-test, * p < 0.05, **p < 0.01.

The REAP method also demonstrated significant increases in the intensity levels of TIA-1 in the nuclear fraction after stress induction as well as a slight increase in the whole cell fraction (Fig. 29A and B). For SFPQ, a moderate increase was also evident in both the nuclear and whole cell fractions (Fig. 29A and C).

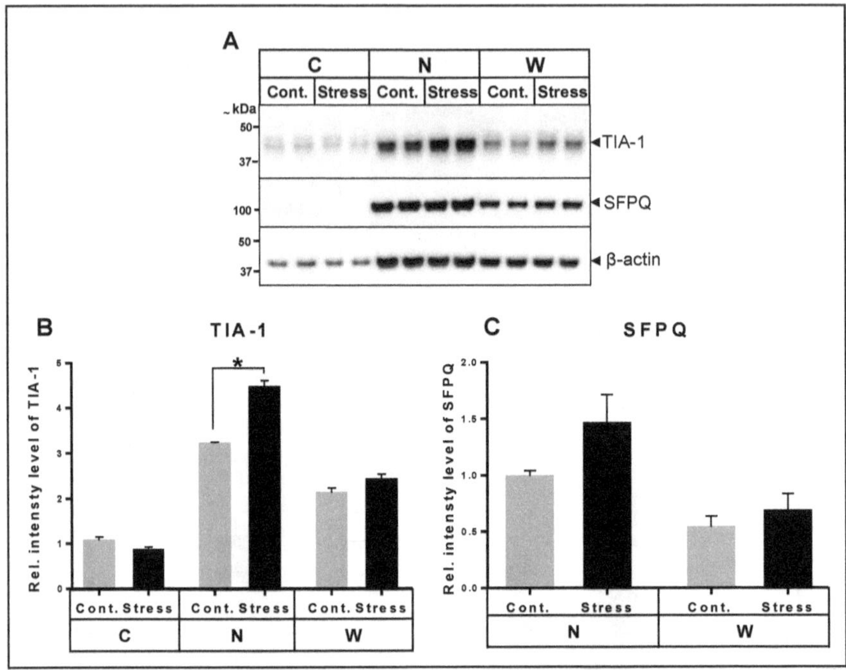

Figure 29: Expression of TIA-1 and SFPQ after subcellular fractionation. A) Stress was induced by treatment with sodium arsenite (0.6 mM, 60 min). Lysates from control (untreated) and stress (arsenite-treated) cells were separated into nuclear and cytoplasmic fractions by REAP method, which were analyzed by immunoblotting with TIA-1 and SFPQ antibodies. Protein amounts were normalized to β-actin. Subcellular fractions were abbreviated as **C**: cytoplasmic extract, **N**: nuclear fraction, and **W**: whole cell lysate. **B and C)** The densitometric analysis of TIA-1 and SFPQ showing significantly increased intensity levels for TIA-1 in the nuclear fraction particularly, as were calculated by *t*-test, p-value *p<0.05.

3.3.2 Role of SFPQ in the tau axis

3.3.2.1 SFPQ downregulation induced by human tau expression

From cell staining (Fig. 27A), co-localization was observed between SFPQ and tau in the cytoplasm. Furthermore, Braak stage-dependent reduction in levels of SFPQ in the entorhinal cortex (EC) has been reported in the human brain (Ke *et al.*, 2012). In order to find out a direct interaction of tau on SFPQ, we expressed human wild-type (WT-tau) and mutant tau (P301L-tau) transiently in HeLa cells.

To assess the expression of tau after transfections, immunoblotting analysis was performed with antibodies specific for total tau (tau-5) and phospho-tau (S199) at 24- and 48 hrs post-transfection. Protein extracts from transfected cells showed an increased level of total tau and its phosphorylated form (Fig. 30A). The densitometric analysis revealed no significant differences in the ratio of phospho-tau to tau between WT-tau and P301L-tau transfected cells (Fig. 30B). A trend of increased phosphorylation could be noted for WT-tau-transfected extracts. Interestingly, expression of human tau led to a significant reduction in the levels of SFPQ, as assessed by immunoblotting at 48 hrs post-transfection in WT-tau transfected cells, compared with control (Fig. 30A and C). For P301L-tau expressing cells, a decreasing trend was also observed at the 48 hrs post-transfection. There were no significant changes observed at 24 hrs post-transfection. For TIA-1 protein, no significant changes could be identified under any of these conditions (Fig. 30A and D).

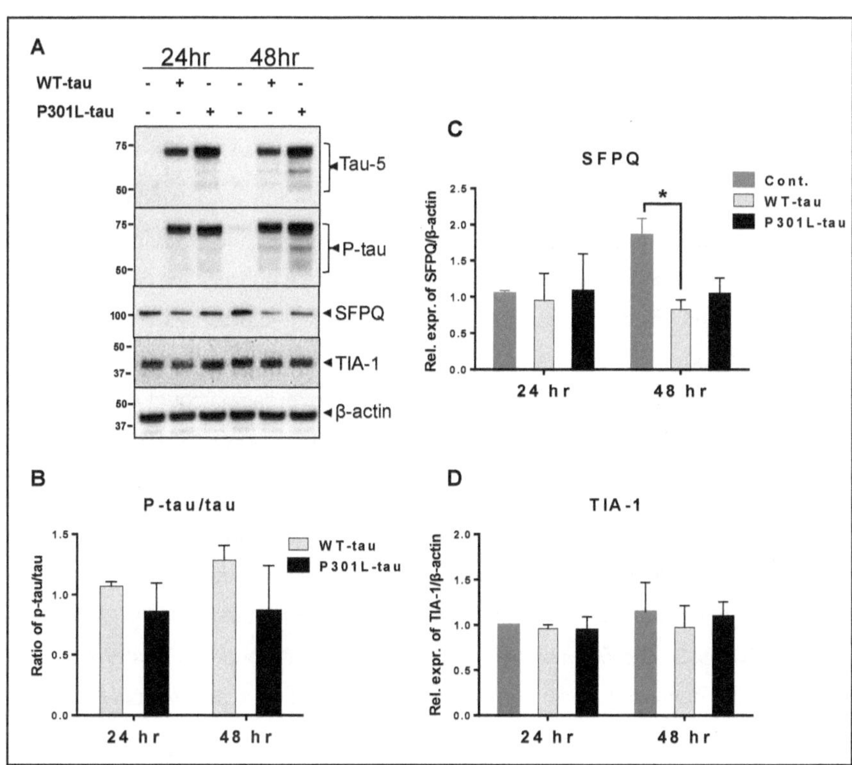

Figure 30: **Tau regulates SFPQ levels in HeLa cells detected by immunoblotting.** A) Representative immunoblots for tau, phospho-tau, SFPQ and TIA-1 after transient transfection of WT-tau or P301L-tau. **B)** Total protein lysates from transfected HeLa cells and control cells were subjected to immunodetection using tau-5 and phospho-tau (S199) antibodies. **C)** The levels of SFPQ were also determined after tau expression. Protein expression was normalized to β-actin. The densitometric analysis revealed significant decrease in the levels of SFPQ in WT-tau-expressing cells at 48 hrs post-transfection. **D)** For TIA-1, no significant differences were observed after transient expression of tau compared with transfection corresponding control. Significance was estimated with one-way ANOVA followed by Tukey post-hoc test for multiple comparisons, *p < 0.05.

Cell viability was significantly reduced after tau expression in both WT- and P301L-tau-expressing cells, with more robust changes in WT-tau-expressing cells at 48 hrs post-transfection (Fig. 44 in annexure data). In summary, immunoblotting analysis indicated that tau has a direct effect on SFPQ in the form of reduction, which coincides well with the massive reduction of SFPQ observed in the rpAD and sCJD patient's brains.

3.3.2.2 Proteomic changes associated with SFPQ downregulation after human tau-expression

Immunoblotting analysis after human tau expression in the cells indicated a direct effect of tau on SFPQ in the form of reduction. To determine the combinatorial effect of tau toxicity and SFPQ reduction in the cells, we used a quantitative SWATH label-free proteomics technology. Quantitative MS-based proteomics has emerged as a powerful technology for investigating mammalian signalling pathways (Huang *et al.*, 2015; Liu *et al.*, 2017). Proteomic alterations after tau expression could provide novel insights into the mechanisms linking tau-pathology to SFPQ dysregulation.

In total, 3597 proteins were identified quantitatively at a critical FDR of 1%. To identify differentially expressed proteins (DEPs), bioinformatic analysis was performed. Proteins with a corrected BH p-value < 0.05 were considered differentially expressed, compared with controls (Fig. 31A and B). Significantly DEPs (314) were divided into two sets: the first consisted of 63 up-regulated proteins, and the second consisted of 251 down-regulated proteins (Tables 16 and 17 in annexure data).

To get deeper insights into the molecular mechanisms associated with significantly modulated proteins, a number of bioinformatic tools were used to examine enriched functional categories and pathways. Firstly, hierarchical clustering and enrichment analysis was performed using Fisher's exact test by Perseus software, in order to have an overview of the significantly enriched functional profile. As shown in Fig. 31C, the significantly regulated proteins were clustered hierarchically across all the samples (columns) and the significantly regulated proteins (rows). Interestingly, the main branch point in the columns separated the tau-expressing cells (Fig. 31C, Lanes 1–12) from controls (mock-transfected or non-transfected controls) (Fig. 31C, Lanes 13–24). No significant differences were observed between WT-tau and P301L-tau expressing phenotypes.

The significantly regulated proteins (rows) were grouped into two main sub-clusters. Cluster 1 included proteins that were upregulated in tau-expressing cells (both WT-tau and P301L-tau). The functional enrichment analysis of cluster 1 by Fisher's exact test indicated highly significant enrichment for substrate specific transporter activity and immune system processes (Fig. 31C: right panel). The downregulated proteins (cluster 2) showed enrichment for functional terms associated with "structural con-stituent of ribosome", and "SRP-dependent co-translational protein targeting to membrane" among others (Fig. 31C: right panel). Additionally, an over-representation analysis was performed using WEB-based Gene SeT AnaLysis Toolkit (WebGestalt). By classifying the 314-DEPs (Liao *et al.*, 2019) based on their "Biological Process", 202 proteins were identified as being related to the metabolic processes, 180 proteins linked to biological regulation, and 139 attributed to be in-volved in response to the stimulus (Fig. 31D). For Molecular Function, 228 proteins were assigned to the protein-binding category, with 109 attributed to be related to nucleic-acid binding, and 107 to ion binding.

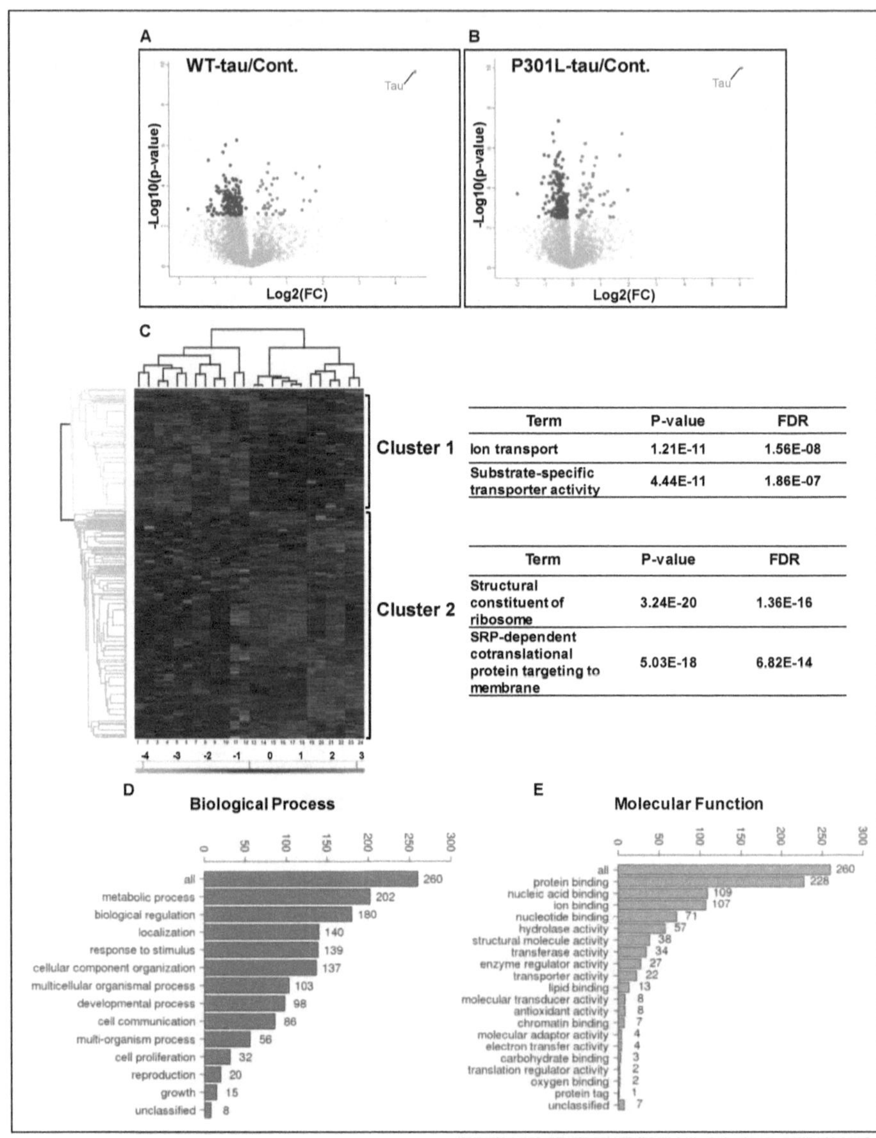

Figure 31: Human tau-expression-induced proteomic alterations: A and B) Volcano plots of pairwise comparisons showing differentially expressed proteins in both WT-tau- and P301L-tau-expressing cells in comparison to controls. **C)** The heatmap of two-dimensional hierarchical clustering analysis of 314 DEPs among technical and biological replicates generated by Perseus software (version 1.5.0.31). The $\log_2$-transformed expression values of significantly regulated proteins were normalized to the Z-score for all the replicates. The columns are representing samples (tau-expressing cells: lanes 1–12 and control: lanes 13–24) and rows indicating significantly modulated proteins (green, down-regulated: red, up-regulated). Functional enrichment analysis of DEPs re-

sulted in multiple categories that were enriched in the two clusters. The enriched functional terms with their p-values and FDR values are shown on the right side (Fisher's exact test with FDR multiple test correction). The Fisher's exact test was used to select the significant GO terms identified by a p-value < 0.05. **D and E)** GO-slim summary from WebGestalt, GO classification of proteins in both clusters. The number of proteins associated with each GO term in the "Biological Process" and "Molecular Function" domains is indicated in red and green bar charts, respectively.

Detailed analysis of the "Molecular Function" GO-terms of DEPs revealed RNA-binding (enrichment ratio = 3.58) and structural constituent of ribosome (enrichment ratio = 7.3) as the most enriched terms (Fig. 32).

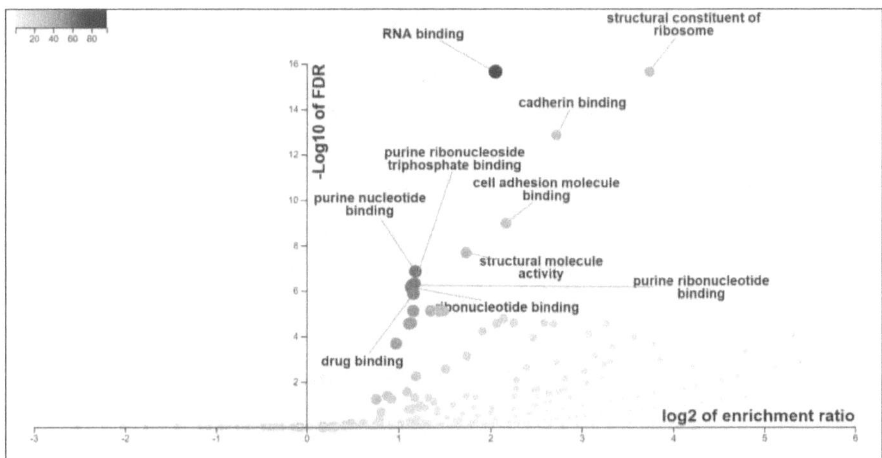

Figure 32: Volcano plot showing significantly enriched Molecular Functional terms from DEPs (314). The vertical axis is showing -Log$_{10}$-FDR value and x-axis represents the log$_2$-transformed enrichment ratio obtained from WebGestalt.

3.3.2.2.1 Canonical pathway analysis

To find out major canonical pathways modulated after human tau expression, Ingenuity Pathway Analysis (IPA, Qiagen, USA) was performed. The proteomic candidates were sorted into 67 canonical pathways. According to their p-values, the top 5 significant pathways identified in WT-tau-expressing cells as compared with mock-transfected cells were eukaryotic initiation factor 2 (eIF2) signalling (p-value = 2.28E-23), regulation of eIF4 and p70S6K signalling (p-value = 6.44E-07), mTOR signalling (p-value = 5.60E-06), mismatch repair for eukaryotes (p-value = 4.93E-05), and pro-line biosynthesis I (p-value = 1.23E-04).

The most significant canonical pathways modulated after P301L-tau expression were EIF2 signalling (p-value = 3.23E-18), interferon signalling (p-value = 2.86E-04), mTOR signalling (p-value = 9.50E-04), DNA double-strand break repair by non-homologous end joining (p-value = 1.11E-03), and telomere extension by telomerase (p-value = 1.28E-03). Overall, EIF2 signalling was the most significant of the en-riched canonical pathways.

3.3.2.2.2 Disease- and function-based protein networks

In addition to pathway mapping, DEPs were also categorized into related diseases and functions. The most significant networks associated with significantly modulated proteins are listed in annexure data for both WT-tau- and P301L-tau-expressing cells, respectively (Tables 18 and 19). The most significant network that was also shared between WT- and P301L-tau-expressing cells was "RNA damage and repair, protein synthesis" (score = 61). Among these networks, two most significant net-works are described below.

3.3.2.2.3 Protein network associated with RNA damage and repair, protein syn-book, cancer

The most enriched network that was common between both types of tau-expressing cells "RNA metabolism and protein synthesis" was composed of 27 focus molecules, including 60S ribosomal subunit, C7orf50, CDK4/6, Eif4g, ERK1/2, IFIT1, importin beta40s subunit, RPL15, RPL18, RPL18A, RPL19, RPL23A, RPL27A, RPL31, RPL32, RPL36, and RPL37A, among others. Biological Processes associated with this network were visualized by Cytoscape (3.6.1) (Fig. 33). Significant Biological Processes associated with this network include large and small ribosomal subunit, RNP complexes, SRP-dependent co-translational protein targeting to membranes, among other processes (Fig. 33).

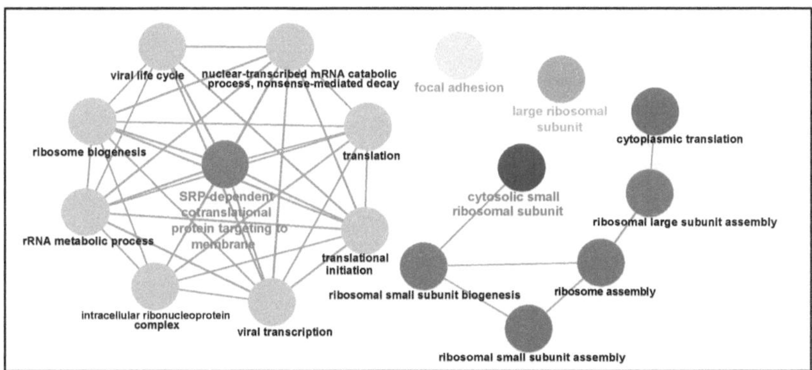

Figure 33: RNA damage and repair, protein synthesis, cancer. The Biological Processes associated with this IPA-based network were visualized by Cytoscape (3.6.1). The Cytoscape plugin ClueGO was used to identify enriched Biological Processes from the network molecules. Nodes with related Biological Processes are marked with the same colour.

3.3.2.2.4 Protein network associated to cell morphology, cellular assembly and organization, DNA replication, recombination, and repair

The second most enriched network was "cell morphology, cellular assembly and organization, DNA replication, recombination, and repair network", consisting of 20 focus molecules from DEPs from proteomic dataset. This complex network was interconnected with Biological Processes associated to homeostatic processes, including telomere maintenance and organization, DNA metabolic processes and chromosome organization, among others (Fig. 34).

Proteomics investigation with functional characterization has provided a comprehensive overview of major alterations associated with tau toxicity and downregulation of SFPQ. Analysis of global proteomic alterations revealed two major themes "RNA metabolism and protein synthesis", and "DNA homeostasis-related processes", that were significantly altered after human tau expression.

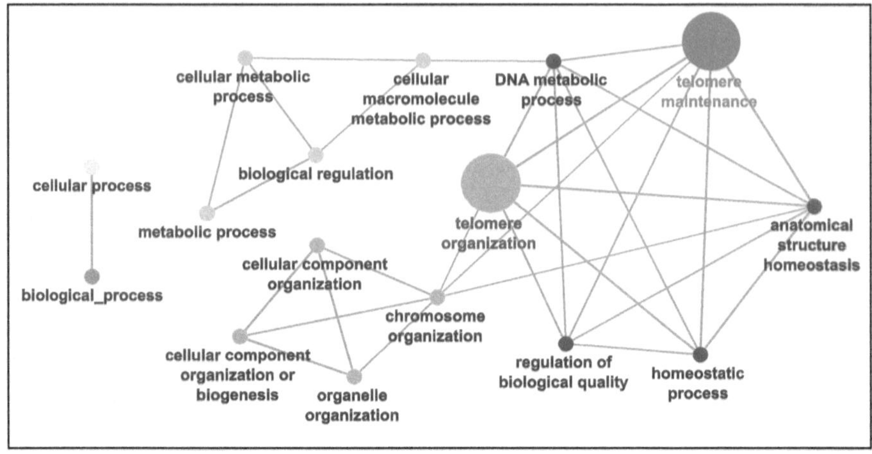

Figure 34: Biological Processes associated with most enriched network molecules from IPA-based network "Cell morphology, cellular assembly and organization, DNA replication, recombination, and repair". The final network was visualized using Cytoscape (3.6.1). The Cytoscape plugin ClueGO was used to visualize Biological Processes from the most enriched network molecules. The nodes from related functional terms are of the same color. The size of the node is corresponding to BH-corrected p-value and width of lines is representing extent of the overlap between related terms.

3.4 Translation of SFPQ-tau-TIA-1 in the 3xTg-AD mice model

Given the long incubation periods of clinically silent neurodegeneration and the manifestation of AD at later stages, knowledge of the modifiable risk factors at presymptomatic stages of the disease is crucial. To extend our results from the analysis of terminal stage pathology from the human brain, a time-dependent expression profile of proteomic signatures was examined in a mouse model of AD, termed 3xTg-AD, both at pre-symptomatic and symptomatic stages of the disease.

Mice were injected with 10% brain homogenate from AD patients. Animals were sacrificed at four time points spanning from early and late pre-symptomatic to early symptomatic and late symptomatic stages, e.g. at 3-, 6-, 9- and 12-months post inoculation (mpi). Total protein expression levels between inoculated and non-inoculated control mice were determined at each time point. In summary, we found that protein expression during disease progression was dynamic and characterized by distinct changes in the global expression levels.

3.4.1 Differential expression of tau in AD mice

Hyperphosphoylation of tau protein and concomitant formation of NFTs is a major hallmark of AD and closely correlates to cognitive loss. Therefore, tau phosphorylation status was investigated in the mice model at all stages of the disease using immunoblotting analysis. The level of total tau was significantly decreased at the terminal stage between the experimental and control groups ($p < 0.05$) (Fig. 35A and B). There was a trend of increased phospho-tau levels at the early pre-symptomatic stage in experimental animals, when compared to controls, with no significant changes observed at later stages (Fig. 35A and C).

3.4.2 Dysregulation of SFPQ at early and late-symptomatic disease stages

Interestingly, levels of SFPQ were significantly altered at early pre-symptomatic phase with a significant increase in comparison to controls (Fig. 36A and D). At middle stages of the disease, expression levels returned to basal level and underwent a direct reversal at late-symptomatic stage of the disease with a drastic reduction at 12 mpi (Fig. 36A and D). Within the sensitivity of the Western blot, the band for SFPQ disappeared completely at the late-symptomatic stage of the disease, which coincided well with the massive reduction of SFPQ at the terminal stage of the disease observed in the postmortem brain, particularly in the rapidly progressive forms of dementia (rpAD and sCJD).

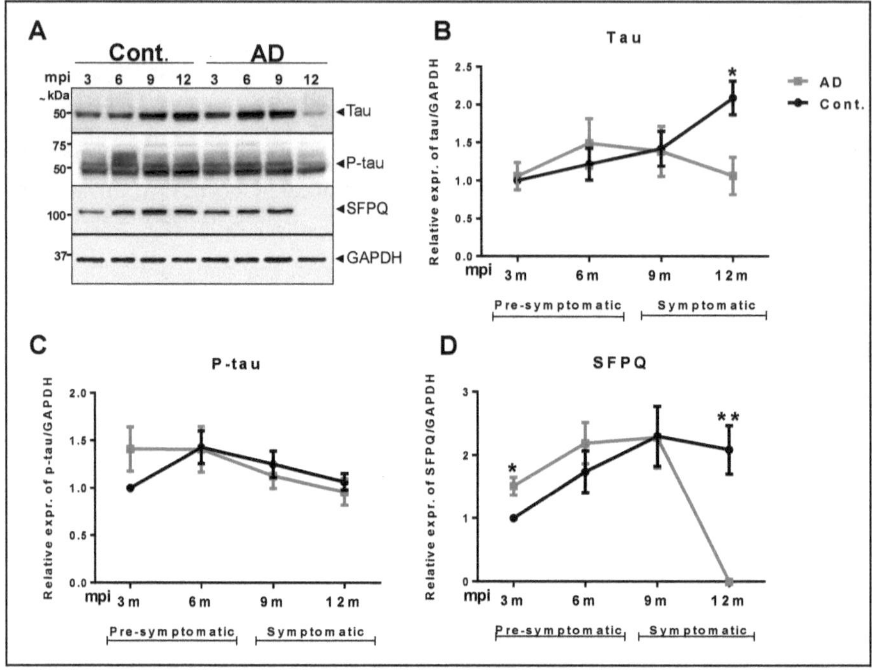

Figure 35: Temporal expression profile of tau, phospho-tau and SFPQ. A) Representative immunoblot images showing expression of tau, phospho-tau and SFPQ from AD (n = 4) mouse brain cortical tissues at the indicated ages (mpi: months post inoculation) and respective controls (n = 4). The GAPDH was used as a loading control. **B-D)** The densitometric analyses from three independent experiments were performed with Image Lab software. Unpaired *t*-test was performed to calculate statistical significance at each time point. *p < 0.05, **p < 0.01, ***p < 0.001.

3.4.3 Alterations in TIA-1 levels at early pre-symptomatic and terminal stage of the disease

Furthermore, we also studied the expression profile of TIA-1 in the AD mouse model at different time points. Two antibodies were used to detect both the C- and N-terminus of TIA-1 (TIA-1 abcam: recognizing amino acids 350 at the C-terminus; TIA-1 sc-166247 recognizing amino acids 37-65 at the N-terminus). The protein levels of TIA-1 were significantly decreased at terminal stages of the disease (12 mpi) for both antibodies (Fig. 36A, B and C). For TIA-1 (recognizing amino acids 350 to the C-terminus), a significant increase was observed at the early pre-symptomatic stage of the disease, which then returned to basal levels at the middle stages of the disease (Fig. 36A and B). For C-terminal specific antibody of TIA-1, a low molecular weight band at the terminal stage (12 mpi) was observed, which seems to be a cleavage product of TIA-1 (Fig. 36A, denoted with star*). The levels of VCP were also assessed at both pre-symptomatic and symptomatic stages in 3xTg-AD mice model. An increase in VCP levels was observed at the terminal stage of the disease, although that was found not to be significant (Fig. 36A and D).

In summary, the time-dependent expression profile with prominent changes in SFPQ, tau and TIA-1 at early pre-symptomatic stages suggested an altered regulation of these molecules at the very early stages of the disease.

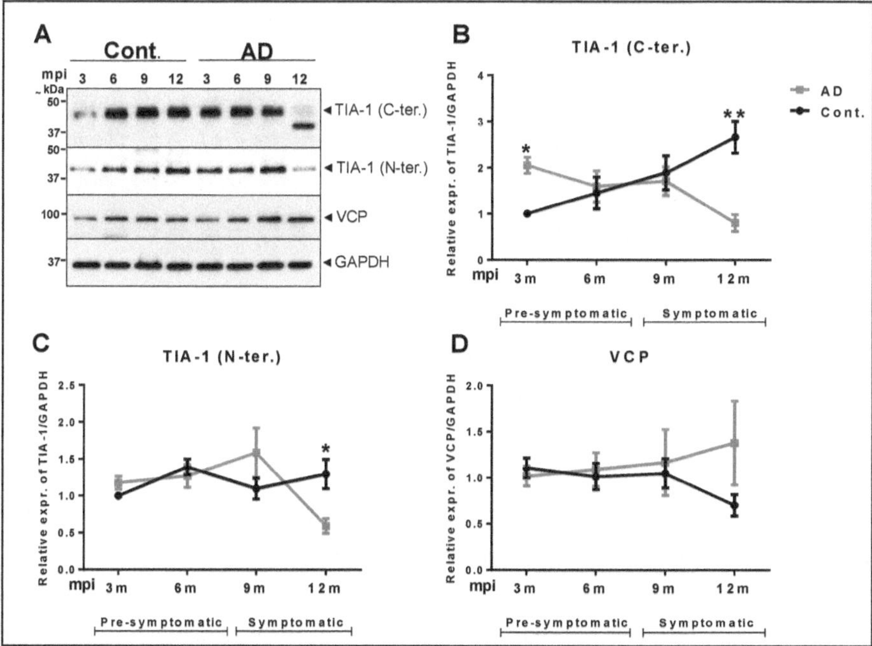

Figure 36: Alterations in TIA-1 levels during disease progression. **A)** Representative immunoblot images showing expression of TIA-1 (C- and N-terminal) and VCP from AD-like (n = 4) mouse brain cortical tissues at the indicated ages (mpi: months post inoculation) and respective controls (n = 4). (*) Band may indicate proteolytical degradation of the C-terminus. The GAPDH was used as a loading control. **B-D)** Unpaired *t*-test was used to calculate significance. *p < 0.05, **p < 0.01.

4 Discussion

Dysfunctional regulation of RNA-binding proteins is a characteristic feature of many neurodegenerative diseases. Although precise mechanisms are still not clear, but it is increasingly evidenced that RBP anomalies are linked to neurodegenerative processes and/or accelerating their progression. To this end, the present study identified and characterised the RNA-binding proteome from human brain frontal cortex of three neurodegenerative entities, namely spAD, rpAD, and sCJD (MM1 and VV2 subtypes), as well as control subjects, using a brain-derived RNA-based pull-down approach followed by mass spectrometry analysis. Cortical region was studied due to maximum pathological manifestations in this region at end stages of the disease (Braak and Braak, 1991). Two mass spectrometry approaches (LFQ-MS and SWATH-MS) were employed in the present study, to gain a deeper insight into the RNA-binding proteome. Proteomic investigation revealed characteristic quantitative and qualitative changes in the identified RNA-binding proteome in a disease subtype-specific manner.

4.1 RBPome alterations in neurodegenerative diseases

Interestingly, global enrichment profile of RNA-binding proteomic candidates from the LFQ-MS analysis demonstrated major similarities between rpAD and sCJD, compared with spAD (Fig.12B). Although rpAD has same core pathological features (Aβ and tau tangles) in common with spAD based on RBPome signatures, rpAD displayed more similarity with sCJD group in comparison with spAD. These findings suggest that, both forms of rapidly progressive dementias (rpAD and sCJD) have similarities in RBP-mediated neuropathogenesis at the terminal stage of the disease. Furthermore, these similarities also argue that these RBP-related processes may have a pathophysiological role in the rapid progression of these diseases.

The differential enrichment analysis of identified RNA-binding proteome revealed distinct proteomic clusters specific to each group (Fig. 12B). Overall, specific enrichment of these distinct protein clusters may indicate differences in the RBP-RNA interactions in response to disease stress. These specific sets of proteins can be of potential relevance to identify subtype-specific differences of these heterogenous diseases. Furthermore, the identification of RNA-binding proteome opens many new therapeutic targets for exploration (Table 15).

As the RBPome changes were more prominent in the rpAD and sCJD groups, we employed quantitative SWATH-MS technology to get a deeper insight into the RBPome of these rapidly progressive forms of dementia. This finding also served as a confirmatory approach to the LFQ-mass spectrometry. It is notable that the RBPome from SWATH-MS virtually recapitulated the LFQ-based proteomic data. In consistence with deep proteome coverage by SWATH-MS, this analysis indicated 477 proteins which showed differences in the abundance between rpAD and two subtypes of sCJD MM1 and VV2. One interesting finding evident from the hierarchical clustering analysis of differentially abundant proteins was the similarity between rpAD and sCJD-MM1 subtype compared with the profile from sCJD-VV2 subtype. Both the rpAD and sCJD-MM1-subtype were segregated into a single cluster, while sCJD-VV2 subtype segregated into a separate cluster. The methionine homozygosity (M/M) at codon 129 of the prion protein gene (*PRNP*) has been described as a risk factor for AD (Gacia *et al.*, 2006; Schmidt *et al.*, 2010). Similarities between the rpAD and sCJD-MM1 subtype observed in the current study support the notion that methionine homozygosity is not only a risk factor for AD, but it may have a role in the rapid progression of the disease.

4.1.1 Functional analysis of MS results

Functional enrichment analysis of the identified RBPome, highlighted major changes related to RNA metabolism, stress response, metabolic processes, and immune system among others. The enrichment of metabolic proteins in the current RBPome dataset demonstrates an interaction between RNA-related processes and metabolic changes during the disease. The regulation of metabolic processes resulting from disease stress is necessary to facilitate survival of the neurons. The RBPs may be engaged in regulatory plasticity, which is required for the instant stress response. In addition, the identification of immune-system-related proteins, supports the integral role of inflammatory processes in the pathogenesis of these neurodegenerative diseases. Finally, our data is also consistent with the well-known functions of RBPs in the post-transcriptional stress response and RNA metabolism. In summary, our findings denote that changes in the RBPs, in addition to metabolic and inflammatory responses, are an integral part of the pathological features of these neurodegenerative diseases.

Furthermore, the Biological Process related to "Localization" including protein and macromolecular localization was a leading term enriched in rapidly progressive AD, indicating disturbances in the protein localization as a pathomechanism in rpAD. Our study identified numerous important proteins, e.g. importin subunit-beta1 (nuclear factor p97), clathrin coat assembly protein AP180 or secretogranin, which were uniquely present in the rpAD dataset. These signatures can be of potential relevance to distinguish the rapid variant of AD from both normal and classical AD-related dementia. Some of these proteins might represent novel biomarkers whose expression, aggregation or posttranslational modifications might reflect distinct elements of the disease process.

4.1.1.1 Proteins involved in RNA metabolism and stress response

The functional categories related to stress response, antioxidant activity, and RNA-binding were specifically enriched in rpAD and sCJD groups, compared with control and spAD groups, indicating a more aggressive dysregulation of RBP-related processes in these fast-progressive forms of dementia. Stress in almost every form has been linked to fast progression in not only AD but also other neurodegenerative diseases, e.g. Parkinson's and Huntington's diseases (Hiller *et al.*, 2017). Studies from many groups and different animal models have reported that stress is a factor leading to fast progression of AD, including increased plaque assembly, hyperphosphorylation of tau, and tangle formation. Not only in animal models, but many studies in humans have elaborated that exposure to extreme stress could also lead to rapid progression of sporadic AD (Herbert and Lucassen, 2016; Mejía *et al.*, 2003), and earlier onset in familial AD (Mejía *et al.*, 2003). Data showing the role of stress response in prion diseases have also been reported (Goggin *et al.*, 2008; Mays *et al.*, 2019).

In summary, data from this study and reports from the literature highlight a significant role of RBPs through stress response in the rapid progression of the disease. Recently, alteration of tau protein has also been linked to stress response in AD (Brunello *et al.*, 2016; Vanderweyde *et al.*, 2016), further strengthening the hypothesis that stress and post-transcriptional regulatory processes associated with RBPs are an integral feature of these neurodegenerative disorders.

4.1.2 Canonical and putative RNA-binding candidates

The present study identified many putative RNA-binding protein candidates (e.g. metabolic and catalytic enzymes) in addition to canonical RBPs, in agreement with previous reports (Shchepachev *et al.*, 2019). The identification of these putative RBPs in the RNA-binding proteome data points towards a cross-talk between RBPs and other proteins (e.g. metabolic enzymes) to meet the everchanging microenvironment surrounding the neuron. Dysregulation of these intricate networks of proteins in the neurons may start a cascade of aberrant signaling, eventually leading to neurodegeneration.

Many studies identifying RNA-binding proteomes have reported a large number of putative RBP candidates, that have no prior linkage to RNA-related functions (Baltz *et al.*, 2012; Beckmann *et al.*, 2015; Castello *et al.*, 2012; Castello *et al.*, 2013). These putative RBPs, which were categorized as enigmRBPs, have many diverse roles in cellular homeostasis, including actin remodeling, protein folding, and many metabolic enzymatic activities; however the functions of most of these enigmRBPs are largely unknown. To date, the role of only a few enigmRBPs, e.g. metabolic enzyme (IRP1), has been discovered (Castello *et al.*, 2015; Hentze *et al.*, 2018). It is notable that some of the identified putative RBP candidates (e.g. SNG3, HEBP1, SV2A) may not have RNA-binding activity themselves. They were identified in the present study due to binding with other proteins that do have RNA-binding activity.

These putative RBP candidates can potentially be beneficial for the cell. Interactions between RNA and these putative RBPs (specifically metabolic enzymes) can be used for spatial sorting of related enzymes. This spatial sorting helps the cell to boost the metabolic flux (Castello *et al.*, 2015).

4.1.3 Prion-like-domain (PLD)-containing proteins

Prion-like-domain-containing proteins can be both beneficial and harmful for the cell (Sabate *et al.*, 2015a; Sabate *et al.*, 2015b). In the current study, twenty-four PLD-containing proteins, fulfilling the requirements to potentially behave as prion-like proteins were identified (e.g. BSN, SFPQ, EWS, PRIO). Prion-like domains are essential for RBP functions and enable them to undergo liquid-liquid phase separation, that is the basis of the formation of higher-order structures, including oligomers and granules (Boeynaems *et al.*, 2018; Riback *et al.*, 2017).

In the present study, we observed that the PLD-containing protein SFPQ is exclusively enriched in the RBPome of rpAD and sCJD, suggesting a potential role of SFPQ in rapidly progressive dementias. Furthermore, SFPQ exhibited a considerably high score for prion-like domain (PLD-score = 28), which is a very crucial factor contributing to pathophysiological functions of PLD-containing proteins. Indeed, we discovered dysregulation of SFPQ in association with tau and TIA-1 proteins, particularly in the postmortem brains from rpAD and sCJD patients.

4.2 Pathological characterization of SFPQ in the human brain

4.2.1 SFPQ dysregulation in the rpAD and sCJD brains

Emerging evidence supports that neurodegenerative anomalies modulate the expression of RBPs (Conlon and Manley, 2017). Interestingly, we identified a significant reduction in SFPQ at the protein level in the frontal cortical region of rpAD and the subtypes of sCJD MM1 and VV2. A trend was also observed for spAD cases, although this was not significant. Ke et al. (2012) reported a significant reduction in SFPQ levels with advanced Braak stages in the entorhinal cortex of AD patients. Specific reduction of SFPQ in both forms of rapidly progressive dementias (rpAD and sCJD) further implies that SFPQ may be involved in the rapid progression of these neurodegenerative diseases. To the best of our knowledge, this is the first study demonstrating dysregulation of SFPQ at both protein and mRNA level in the frontal cortex of specifically rpAD subjects.

The reduction of SFPQ has been linked to behavioral anomalies in mice, neuronal loss, and phospho-tau accumulation (Ishigaki *et al.*, 2017). In another study, loss of SFPQ was found to lead to apoptosis in zebra fish, linking downregulation of SFPQ to neuronal cell death (Lowery *et al.*, 2007). Splicing factor proline and glutamine rich is a predominantly nuclear protein involved in multiple functions in the neurons, including transcription, alternative splicing, DNA damage and repair, and transport of mRNAs through long axons (Knott, *et al.*, 2016; Yarosh *et al.*, 2015). It can be concluded that downregulation of SFPQ can contribute to neurodegeneration by affecting multiple functions of SFPQ, specifically in the nucleus.

At the mRNA level, SFPQ expression was elevated in rpAD, in contrast to the reduction at protein level. One plausible explanation for this observation could be that transcription of SFPQ is increased in order to compensate for loss of SFPQ at the protein level. The increased mRNA levels of SFPQ could also contribute directly to neurodegeneration. The excessive mRNA may sequester many proteins necessary for other cellular signalling, as has been noted for some other proteins involved in neurodegenerative disease (Greco *et al.,* 2006; Sellier *et al.,* 2014; Tassone *et al.,* 2004). These aberrant masses of mRNA and protein may convert into the inclusion bodies (Greco *et al.,* 2006; Iwahashi *et al.,* 2006).

4.2.2 SFPQ dislocation in the brain of rpAD patients

Cytoplasmic mislocalization of many nuclear factors has been defined as a pathomechanism in several neurological diseases (Barmada *et al.,* 2010; Bishof *et al.,* 2018; Neumann *et al.,* 2006; Vance *et al.,* 2013). Additionally, in the current study, a drastic nuclear depletion was detected for SFPQ in the frontal cortex of rpAD subjects. Previously, one study has reported a complete dislocation of SFPQ in the hippocampus of AD patients (Ke *et al.,* 2012). Specifically, the higher dislocation/depletion rate (91% of cells) observed in rpAD cases in the current study as compared to spAD (51%) and controls (43%) suggests an important role of SFPQ in the rapid progression of the disease. Lu et al. (2018) demonstrated a moderate dislocation of SFPQ in AD brains, which was consistent with the observations regarding spAD described in this study. Based on data from the current study, from the frontal cortex and literature reports (showing dislocation in hippocampus), it can be concluded that dislocation of SFPQ is an important feature of Alzheimer's pathology, and different brain regions have variable intensity of SFPQ dislocation. Increased dislocation in rpAD subjects may be due to higher cell death (through apoptosis), as cytoplasmic localization of SFPQ has been observed in apoptotic cells. The nuclear depletion of SFPQ, particularly in rpAD, may contribute to neurodegeneration by both the loss of nuclear functions (Lu *et al.,* 2018) and toxic functions in the cytoplasm.

Nuclear loss of SFPQ could render cells more prone to DNA damage, considering that solid evidence has been established for a crucial role of SFPQ in the DNA double-strand break repair processes (Jaafar *et al.*, 2017). Lu et al. (2018) reported a disrupted DNA organization in association with SFPQ dislocation. SFPQ depletion from the nucleus may induce cell death by contributing to mitosis, as a redistribution in the cytoplasm was observed in mitotic cells (Shav-Tal *et al.*, 2001).

4.2.3 SFPQ co-localization with the SG marker TIA-1 in the rpAD brain

Nuclear depletion and dislocation of SFPQ was concomitantly associated with its cytoplasmic co-localization with TIA-1 (a classical marker of SGs) in rpAD subjects. A ring-shaped SFPQ was observed around the nuclei which co-localized with TIA-1. Previously, pathological SGs have been linked to mislocalization of tau, FUS and TDP-43 (Bosco *et al.*, 2010; Liu-Yesucevitz *et al.*, 2010; Vanderweyde *et al.*, 2016; Yasuda *et al.*, 2017).

Furthermore, SFPQ translocated from the nucleus and assembled into cytoplasmic TIA-1-positive stress granules in response to oxidative stress conditions in cultured cells. These findings from the current study indicate a role of SFPQ in the stress response under physiological conditions. The identification of SFPQ in the stress granule-interactome of U2OS cells (Jain *et al.*, 2016) further confirms the involvement of SFPQ in the stress response. Pathological and persistent TIA-1-positive-stress granules have been implicated in AD (Apicco *et al.*, 2018). Increased intron retention in *SFPQ* transcript has been reported in ALS patients (Luisier *et al.*, 2018). Extensive binding of SFPQ to its retained introns leads to higher cytoplasmic abundance. This corroborates with our findings of increased mRNA levels of SFPQ in the rpAD brain, which might contribute to the altered localization observed in the rpAD cases.

Though predominantly a nuclear protein, multiple localizations of SFPQ have been reported. Phosphorylation at C-terminal tyrosines leads to accumulation at the nuclear envelope or in the cytoplasm (Lukong *et al.*, 2009; Otto *et al.*, 2001). This redistribution of SFPQ has been linked to cell cycle arrest (Lukong *et al.*, 2009), apoptosis (Galietta *et al.*, 2007; Shav-Tal *et al.*, 2001), and splicing abnormalities (Heyd and Lynch, 2010; Melton *et al.*, 2007). Furthermore, a role for SFPQ in the IRES (internal ribosome entry site)-mediated translation in the cytoplasm has also been reported

(Sharathchandra *et al.*, 2012). From our results and literature reports, we propose three possible mechanisms responsible for the dislocation of SFPQ from the nucleus: (1) Chronic stress may turn physiological SGs into pathological, and hijack SFPQ in the cytoplasm. (2) Retained introns in combination with increased levels of SFPQ transcript may sequester SFPQ in the cytoplasm. (3) Redistribution of SFPQ into the cytoplasm is linked to cell death (apoptosis), specifically in rpAD cases.

4.2.4 SFPQ and neurofibrillary tangles

Co-aggregation of some splicing factors with tau protein in the cytoplasmic inclusions has been reported for both sporadic and familial AD cases (Bai *et al.*, 2013; Bishof *et al.*, 2018; Diner *et al.*, 2014). The present study identified a depletion of both SFPQ and phospho-tau from the nucleus in brain tissue from rpAD patients. Both proteins showed cytoplasmic/perinuclear co-localization as compared to nuclear co-localization in control cases. One possible explanation for this observation could be that, at earlier stages, activated kinases can phosphorylate not only tau protein but also SFPQ. Cytoplasmic association of SFPQ with phospho-tau tangles, and their complete loss from the nucleus, suggests a change in the function of these proteins, extending the pathogenic role of tau to the nuclear processes.

Hernandez-Ortega et al. (2016) described a complete depletion of nuclear tau in the neurons bearing NFTs in hippocampal-CA1, entorhinal and temporal neocortical regions at the terminal stages of the disease. This altered localization of nuclear tau was associated with chromatin modifications (Hernandez-Ortega *et al.*, 2016). In rpAD cases showing complete loss of nuclear tau, DNA-protective role of nuclear tau (Sultan *et al.*, 2011) or heterochromatin stabilization function (Hernandez-Ortega *et al.*, 2016; Sjöberg *et al.*, 2006) will be disturbed. Furthermore, SFPQ has also been linked to telomere maintenance (Petti *et al.*, 2019). Its dislocation from the nucleus along with tau protein may lead to impaired DNA functions and results in aberrant gene regulation. In summary, our data in corroboration with previous reports suggest the involvement of nuclear tau, possibly in conjunction with SFPQ, in the pathology of rpAD.

Interestingly, we detected a predominant nuclear localization for phospho-tau (S199) in the control subjects. Nuclear localization of phospho-tau has been reported in normal cell lines (Shea and Cressman, 1998), mouse brain (Lambert *et al.*, 1995; Lu *et al.*, 2013), and human brain (Brady *et al.*, 1995). However, the significance of phospho-tau in the nucleus under control conditions is not clear. Localization of phospho-tau in the nucleus under control conditions, observed in the present study in combination with literature reports, highlights a potential role of tau in genome surveillance. Further knowledge on nuclear and, in particular, phosphorylated tau may provide clues in understanding pathological features of nuclear tau in neurodegenerative diseases.

In spAD subjects, tangled tau was even observed in the nucleus in agreement with previous findings (Fernandez-Nogales *et al.*, 2017; Lu *et al.*, 2018), where it co-localized with SFPQ. Identification of tangled tau in the nucleus extends pathological features of tau not only to axon or somatodendritic compartments but also to nuclear processes. Previously, rod-like deposits of tau were also identified in the nuclei of subjects with AD and Huntington's disease (Lu *et al.*, 2014). Collectively, evidence from the current study in corroboration with previous studies suggests a role for tau within the nucleus under normal and disease conditions. Furthermore, dislocation/depletion of both SFPQ and phospho-tau, specifically in the rpAD cases, may be linked to distinct molecular pathways in subtypes of AD.

Biochemically, there were no significant differences detected for total tau in the current study in the postmortem brains. We were able to detect significant differences in the SDS-resistant HMW-tau (~120KDa) between spAD and control subjects. Interestingly, a trend for lower ratio of phospho-tau/tau was observed in rpAD cases as compared with spAD patients, in line with a previous report (Ba *et al.*, 2017). This decreased phospho-tau/tau ratio indicates a reduced rate of tau phosphorylation in rpAD subjects, contrary to higher phosphorylation, which is a cardinal feature of spAD pathology. This difference in the ratio of phospho-tau/tau may be of potential significance for the development of a progressive form of AD.

4.2.5 SFPQ co-localizes with oligomeric tau in the rpAD

Although cytoplasmic tau deposits are a burden for the cell, it is rather the toxic soluble oligomeric species of tau that are the real culprits associated with cognitive decline, neuronal dysfunction, and death (Guerrero-Muñoz *et al.*, 2015; Shafiei *et al.*, 2017). Interestingly, significant co-localization was detected for tau oligomers and

SFPQ in the rpAD subjects. This interaction between SFPQ and tau oligomers has potential relevance for oligomerization, and subsequent misfolding of tau protein. Previously, co-aggregation of tau oligomers with TIA-1 has been reported in animal models of tauopathy (Apicco *et al.*, 2018; Vanderweyde *et al.*, 2016). In response to different kinds of stresses, SG formation brings together many intrinsically aggregated-prone proteins (e.g. TIA-1, tau, and SFPQ) to form reversible SGs. Chronic stress can turn these physiological SGs to insoluble and pathological SGs (Wolozin, 2012). Based on results from the current study, demonstrating

A) stress-induced redistribution of SFPQ into cytoplasmic TIA-1-positive SGs in cultured cells,
B) its nuclear depletion and co-localization with cytoplasmic TIA-1 in the human brain of rpAD subjects,
C) SFPQ co-localization with tau oligomers and tangles in the rpAD brains,
D) its high score for PLD (PLD score = 28), and
E) its LLPS property (catGRANULES score=1.66),

it is reasonable to propose that SFPQ is an important component of AD pathology, particularly of rapidly progressive AD.

Our findings suggest that dysregulated SFPQ can interact with (hyperphosphorylated) tau and its oligomeric form in stress granules, providing an intimate link between SFPQ, tau pathology and SGs. Dislocation of SFPQ may lead to impaired nuclear functions of SFPQ, e.g. DNA damage and repair, telomere stability, and splicing abnormalities. Toxic gain of function in the cytoplasm may contribute to aberrant dynamics of SGs, oligomerization, and misfolding of tau protein (Fig. 37). Based on results from the current study, we could state that SFPQ pathology is linked with tau protein, more robustly in rapid progressive form of AD. These results provide a new insight into the relationship between SFPQ and tau pathological features in rpAD.

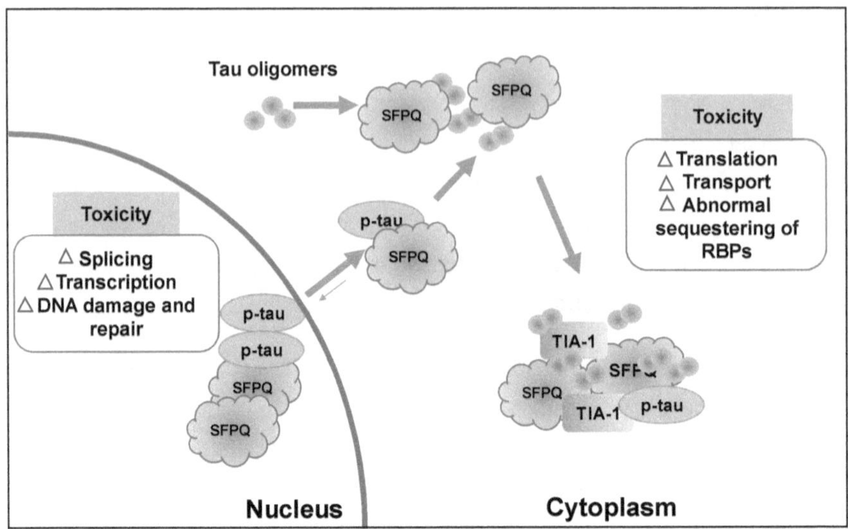

Figure 37: Pathological characteristics of SFPQ, phospho-tau and TIA-1 in the human brain. This figure describes pathological features observed for SFPQ, TIA-1 and phospho-tau in the postmortem brains of rpAD cases. The RNA-binding protein SFPQ is normally localized predominantly in the nucleus, where it performs multiple functions including splicing, transcription, and DNA damage and repair. In the control subjects, significant co-localization was observed for SFPQ and phospho-tau (S199) in the nucleus (1). In rpAD subjects, a complete nuclear depletion and cytoplasmic co-localization was observed for SFPQ and tau tangles (2), and tau oligomers (3). Furthermore, SFPQ co-localization with TIA-1 was identified significantly in brain samples from rpAD patients (4). At protein level, SFPQ was significantly downregulated, and TIA-1 was upregulated in rpAD group, in comparison to spAD group. Our findings suggest that dysregulated SFPQ can interact with (hyperphosphorylated) tau and its oligomeric form in stress granules, providing an intimate link between SFPQ, tau pathology and SGs. Overall, dysregulation of SFPQ in the form of reduction and nuclear depletion/dislocation can affect the cell biology by two-ways. 1) Dislocation of SFPQ may lead to impairment of nuclear functions of SFPQ e.g. DNA damage and repair, telomere stability, and splicing abnormalities. 2) Toxic gain of function in the cytoplasm may contribute to aberrant dynamics of SGs as well as oligomerization and misfolding of tau protein.

4.2.6 Differential regulation of SG marker TIA-1 in disease-subtype-specific manner

Previously, TIA-1 has been linked to tau misfolding (Vanderweyde *et al.*, 2016) but the expression of TIA-1 has not been investigated in the various AD entities. In the present study, we detected a significant reduction in TIA-1 levels in spAD and sCJD-subtypes. It was shown that TIA1 reduction (haploinsufficiency) is cytoprotective *in vivo*, leading to increased survival in a mouse model of tauopathy due to reduced neurodegeneration and improvement in cognitive deficits (Apicco *et al.*, 2018). A significant increase in TIA-1 levels in rpAD as compared to spAD may contribute to higher neurodegeneration in rpAD cases.

Our data show that valosin-containing protein is specifically enriched in the RNA-binding proteome from rpAD and sCJD subjects, suggesting a possible involvement of VCP in these diseases. In our study, a significant increase of mRNA coding for VCP was detected in rpAD subjects. This protein plays an important role in the clearance of SGs (Turakhiya *et al.*, 2018). Malfunction of VCP due to mutations has been associated with increased build-up of stress granules in other neurodegenerative diseases, e.g. ALS and fronto-temporal dementia (FTD) (Ramaswami *et al.*, 2013; Wolozin, 2012). Elevated expression of VCP at the mRNA level may have a role in the formation and clearance of neuronal inclusions in various neurodegenerative diseases (Mori *et al.*, 2013). It can be concluded that VCP dysregulation is a common pathological feature of several neurodegenerative diseases including rpAD.

4.3 Translational aspects of SFPQ in cellular models

4.3.1 Cellular model of stress

To understand the role of target proteomic candidates for SG biology, HeLa cells were used as an appropriate model for stress induction (Aulas and Vande Velde, 2015). These cells offer numerous technical advantages, as they are easy to maintain in culture and reliably result in high transfection efficiency (Bali *et al.*, 2012). Though non-neuronal, this cell line was used for the current study because it produced prominent SGs. Compared to other cell lines, the increased size of HeLa cells allows the accurate distinction of cytoplasmic SGs.

4.3.2 SFPQ recruitment into SGs after oxidative stress treatment

In this study, we demonstrate that SFPQ is recruited into SGs that are induced in response to treatment with sodium arsenite. The SFPQ protein is an important component of RNA transport granules in dendrites (Kanai *et al.*, 2004; Kunde *et al.*, 2011; Zhu *et al.*, 2005). However, it is not known whether SFPQ plays a role under stressful conditions. In the current study, it was found that endogenous SFPQ colocalizes with the core SG marker TIA-1 following exposure to oxidative stress. SFPQ is localized predominantly in the nucleus, but treatment with sodium arsenite induced the redistribution of this protein into the cytoplasm. However, the quantity of translocated protein was small relative to the total amount of the protein. This redistribution led to the formation of cytoplasmic inclusions, which co-localized with SG marker, indicating that these inclusions are SGs. Association of SFPQ with SGs can be mediated through protein-protein or RNA-protein interactions. Ke et al. (2012) reported a redistribution and cytoplasmic accumulation of SFPQ in the form of vesicular aggregates in SH-SY5Y cells after overexpression of tau. Cytoplasmic redistribution in the form of inclusions has also been observed in N2a cells after SFPQ overexpression (Lu *et al.*, 2018). Both of these studies reported the formation of cytoplasmic inclusions of SFPQ, but the nature of these inclusions is not known. Colocalization of SFPQ with TIA-1-positive stress granules, as shown in the present study, indicates that these are most likely SGs.

In summary, our results highlight a role of oxidative stress in the cytoplasmic redistribution of SFPQ and its incorporation into SGs. Furthermore, we also found a significant increase in SFPQ intensity levels after stress induction. It was reported that SFPQ sensitizes neurons to excitotoxic damage *in vitro* (Xu *et al.*, 2005). Our findings indicate an important function of SFPQ under stressful conditions, allowing significant flexibility in gene regulation, therby enabling the cell to adjust in response to different environmental conditions. Based on our findings demonstrating oxidative-stress-mediated redistribution of SFPQ into TIA-1-positive cytoplasmic SGs, it can be concluded that chronic stress may lead to nuclear depletion of SFPQ, which was observed in the postmortem brains of rpAD cases.

4.3.3 Tau, TIA-1 and SFPQ in stress granules

In our model of stress induction, we observed a co-localization of tau and phospho-tau with TIA-1. Interestingly, for phospho-tau, we observed both nuclear and cytoplasmic localization with predominant nuclear reactivity, which was increased in response to stress treatment. Although tau is considered predominantly a cytosol-enriched protein, several studies have reported its nuclear localization in both neuronal (Siano *et al.*, 2019, Ulrich *et al.*, 2018; Wang *et al.*, 1993) and non-neuronal cell lines, including HeLa cells (Sjöberg *et al.*, 2006). Higher immunoreactivity for phospho-tau in the nucleus during stress exposure may indicate a role of tau in cytoprotective mechanisms. Previously, a role of nuclear tau has been implicated in the compensatory mechanisms of the cell through epigenetic changes upon stress exposure (Frost *et al.*, 2014; Mastroeni *et al.*, 2011; Sanchez-Mut and Graff, 2015; Sultan *et al.*, 2011). Notably, we found co-localization of SFPQ with tau and phospho-tau in the granules after stress exposure. This co-localization of SFPQ and phospho-tau in the granules *in vitro* as well as in postmortem human brain of rpAD subjects further strengthen the possibility that SFPQ contributes to tau pathology through the formation and/or stabilization of stress granules.

In the present study, immunoblotting analysis of total cell lysates after stress induction also indicated an increase in tau phosphorylation in agreement with immunocytochemical data. Our results suggest that the phosphorylation status and distribution of tau, particularly phospho-tau, are both modified by oxidative stress in HeLa cells. To rule out the possibility of a cell-type-specific increase in tau phosphorylation, we also investigated the phosphorylation status of tau protein in the neuronal cell line SH-SY5Y. A similar increase was observed in tau phosphorylation in SH-SY5Y cells after stress induction. These findings confirm the stress-dependent increase in tau phosphorylation in both neuronal (SH-SY5Y) and non-neuronal (HeLa) cell lines. An increased activity of GSK-3β has been reported in different cell lines leading to hyperphosphorylation of tau after oxidative stress treatment (Feng *et al.*, 2013; Lovell *et al.*, 2004). One of the earliest events occurring in the progression of AD is an elevation in tau phosphorylation (Huang *et al.*, 2016). Our results confirm oxidative-stress-induced phosphorylation and redistribution of phospho-tau into cytoplasmic SGs along with SFPQ, providing a possible mechanism for co-aggregation and mislocalization of both proteins (Fig. 38).

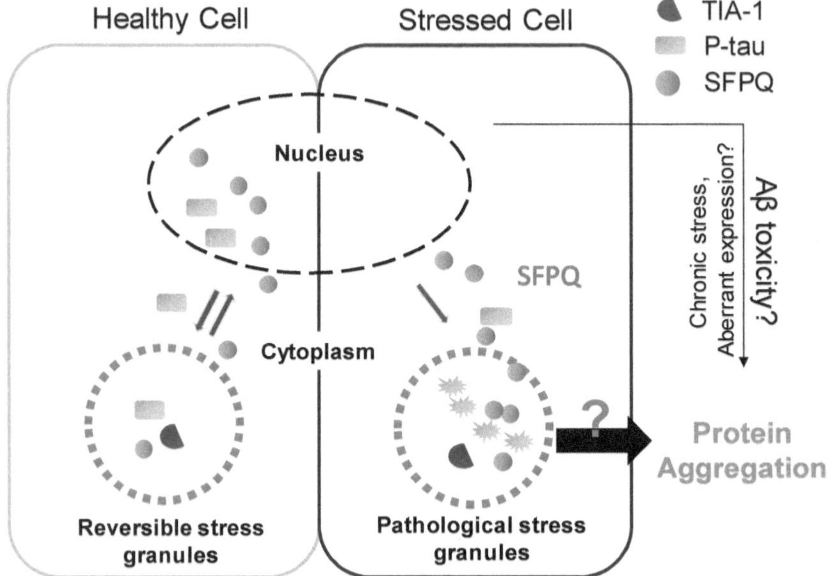

Figure 38: Current working model for SFPQ and tau-pathological features in the rapidly progressive form of Alzheimer's disease. The left box of the picture depicts nucleocytoplasmic translocation of SFPQ and phospho-tau including their assembly into stress granules based on our data from the cellular model of stress. Oxidative-stress-induced redistribution of SFPQ and phospho-tau into the cytoplasm results in the formation of stress granules. These reversible stress granules will be resolved upon removal of the stressors. The right box is depicting pathological features of both proteins observed in the human brain of patients with rpAD at terminal stage of the disease. A complete nuclear depletion of SFPQ and phospho-tau was observed in the postmortem brain of rpAD cases. Furthermore, SFPQ co-localized with tau tangles, tau oligomers and TIA-1 in the cytoplasm in the human brain and in cultured cells, providing a possible mechanism of SFPQ and nuclear tau dislocation through pathological SGs.

4.3.4 Biochemical characterization of stress-granule components

The increase in intensity levels of phospho-tau, TIA-1 and SFPQ during oxidative stress, as observed in this study, raised a concern as to whether this increase was due to increased expression of these proteins or due to consolidation of these proteins into granules. To confirm this, the REAP method (Suzuki *et al.*, 2010) was used, allowing very rapid fractionation (2 min) without the involvement of intermediate, time-consuming lysis steps, thus minimizing the chance of technical processing bias in the assessment of protein expression. The REAP method confirmed that there is an increase in tau phosphorylation after oxidative stress. A significant increase in phospho-tau levels was detected in cytoplasmic and nuclear fractions, with a stronger increase in cytoplasmic fractions. Chronic oxidative stress leads to increased tau phosphorylation in neuronal cultures (Su *et al.*, 2010; Zhu *et al.* 2005).

Our experiments using biochemical fractionation confirmed that oxidative stress mediates an increase in tau phosphorylation and its redistribution into the cytoplasm. However, the amount of translocated phospho-tau into the cytoplasm was small, as compared to the total tau pool. The levels of TIA-1 increased significantly in nuclear fractions and slightly in whole cell lysate fractions. Furthermore, subcellular fractionation also indicated a slight increase in SFPQ levels in both nuclear and whole cell fractions, although that was not significant. One plausible explanation for this observation could be that the increase in SFPQ levels observed in the total cell lysates was not high enough to be detected in the volume-based normalizations used in the REAP method. In summary, these results confirm increased phosphorylation and redistribution of tau into the cytoplasm in response to oxidative stress. The REAP method also confirmed increase in TIA-1 levels after stress treatment.

4.3.5 Role of SFPQ towards tau axis

The Braak-stage-dependent reduction in SFPQ levels in the entorhinal cortex of human brain (Ke *et al.*, 2012) suggests a tau-dependent downregulation of SFPQ. To decipher the role of tau, we expressed recombinant WT or mutant human tau (P301L-tau) in HeLa cells and found a significant increase in total and phosphorylated tau levels after transient transfection. However, there were no significant changes observed in net phosphorylation (phospho-tau/tau) between WT- and P301L tau-expressing cells. Immunoblotting analysis did not show significant differences in the levels of SFPQ 24 hrs post-transfection.

To further extend our knowledge about putative long-term effects of tau expression, we also investigated changes in SFPQ levels 48 hrs post-transfection. Interestingly, immunoblotting analysis revealed a significant reduction in SFPQ levels in WT-tau-expressing cells as compared to controls. These results indicate that expression of tau leads to a reduced SFPQ expression in both WT- and P301L-tau-expressing cells, but a significant downregulation was only found in WT-tau-expressing cells. Downregulation of SFPQ at a transcriptional level has also been reported in pR5 mice expressing P301L-tau (Ke *et al.,* 2012). Given the reduced SFPQ expression in the human brain frontal cortex of rpAD subjects and in transfeced cells overexpressing tau, our data suggest a tau-dependent modulation of SFPQ expression.

The cell viability assay indicated decreased cell viability in HeLa cells after tau-expression as compared to controls. Increased susceptibility to cell death has been reported after WT-tau expression in other cell lines as well e.g. SY5Y cells (Delobel *et al.,* 2003). Similarly, knockout of *whitesnake/sfpq* in zebra-fish has been associated with increased apoptosis (Lowery et al., 2007). Furthermore, our immunoblotting analysis indicated that total levels of TIA-1 were not altered under any of these conditions, suggesting that tau has no effect on the total levels of TIA-1. A role of tau in the regulation of the TIA-1 interactome has been reported previously (Vanderweyde *et al.,* 2016). In summary, our data demonstrate that human WT-tau expression leads to SFPQ reduction and higher susceptibility to cell death.

4.3.5.1 Dysregulated pathways associated with tau-mediated downregulation of SFPQ

To explore pathways and mechanisms associated with tau-dependent downregulation of SFPQ, this study employed a quantitative proteomic technology called SWATH-MS. Combination of three functional enrichment strategies (IPA analysis, Fisher's exact test and WebGeStalt-based enrichment analysis) indicated two major themes that were altered: the first was associated with RNA metabolism (RNA damage and repair and cytoplasmic translation via the eiF2 and eiF4 pathways) and the second as associated with DNA damage and repair. Overall, most of the proteins were downregulated (251) after expression of human tau, as compared with the number of up-regulated proteins (63), suggesting a reduction in global translation. Likewise, most of the proteins belonging to the first theme were downregulated. Given the fundamental importance of protein synthesis machinery in the neurons, it is most likely that tau-mediated aberrant synthesis of ribosomal proteins and subunits is harmful to numerous complex neuronal processes (Rangaraju *et al.,* 2017; Slomnicki *et al.,* 2016). Another possible reason for changes in the synthesis of specific sets of proteins, as observed in this study, might be due to aberrant interaction between tau and TIA1, which could sequester specific mRNAs, change the synthesis of particular sets of proteins and, therefore, contribute to tau pathology (Apicco *et al.,* 2018; Vanderweyde *et al.,* 2016). A link between AD and altered or reduced global translation was recognized initially in 1989 (Langstrom *et al.,* 1989). Significant impairment in ribosomal function has been reported in multiple cortical regions in patients with mild cognitive impairment (MCI) and AD, due to reduced protein synbook, reduced ribosomal and transfer RNA levels (Ding *et al.,* 2005; Hernandez-Ortega *et al.,* 2016). Downregulation of many ribosomal proteins was observed in our data. Recently, impaired synthesis of ribosomal proteins was reported in a tau-transgenic mouse model of FTD (Evans *et al.,* 2019) and in cultured cells (Maina *et al.,* 2018). In summary, our findings confirm impaired protein synthesis as a pathomechanism through which pathological tau can disrupt cellular homeostasis.

The second major theme associated with DEPs was DNA damage and repair. Major proteins belonging to this functional category (XRCC5, XRCC6, FEN1, MSH6, POLD1, PCNA, RFC5) exhibited a reduced expression. A malfunction of these proteins associated with "telomere organization and mismatch-repair in eukaryotes" suggests disturbed DNA-metabolic processes. Tau-mediated DNA disorganization has important implications for AD. These findings confirm a toxic role of tau in the nucleus. Previous investigations in MCI patients have demonstrated a role of DNA-damage in the development of AD and other neurodegenerative diseases (Bucholtz and Demuth, 2013). Based on our results, it is tempting to speculate that tau-induced DNA abnormalities may contribute to AD-related neurological deficits.

There are no reports providing evidence for the involvement of tau in DNA-repair processes (Rossi *et al.,* 2013) or in telomere preservation. A chaperone role for tau protecting genomic DNA against free radicals or heat-induced stress damage has been recently reported (Sultan *et al.,* 2011; Wei *et al.,* 2008), as well as functions in chromatin stabilization (Rossi *et al.,* 2008). Disturbances observed in telomere maintenance and organization could be attributed to a reduction of SFPQ levels, as recently established for SFPQ in the regulation of telomere integrity (Petti *et al.,* 2019).

In summary, the reduced SFPQ levels after tau expression *in vitro*, in the human brain of rpAD and sCJD subjects, and in the 3xTg-AD animals suggest that this depletion may lead to impaired nuclear DNA functions. As observed in the frontal cortex in this study and in the hippocampus by others (Ke *et al.,* 2012), mislocalization of SFPQ suggests a disturbance in multiple roles including DNA damage, transcription, alternative splicing, and transport machinery, eventually contributing to neurodegeneration.

4.4 Dysregulation of SFPQ, tau, and TIA-1 in 3xTg-AD mice

To address the physiological significance of reduced expression of SFPQ in the postmortem human brain, the time-dependent expression profile of proteomic signatures was investigated in the model of 3xTg mice both at pre-symptomatic and symptomatic stages of the AD-like disease. The 3xTg-AD mouse model has three mutations (APP Swedish, MAPT P301L, and PSEN1 M146V) in the brain (Oddo *et al.*, 2003) and develops Aβ-plaques and tau tangles in the hippocampus from the age of 6 months (Belfiore *et al.*, 2019).

In the current study, we identified a significant reduction in the levels of total tau at late-symptomatic stage of the disease in 3xTg-AD mice as compared to controls. No significant changes were observed for phospho-tau (S199), which was consistent with previous findings in 3xTg-AD mice (*Li et al.*, 2019). Furthermore, we uncovered that SFPQ and TIA-1 were already significantly elevated at the early pre-symptomatic phase (3 mpi) of the disease, which indicates their relevance as early disease-modifying targets in 3xTg-AD mice (Fig. 39). This upregulation in 3xTg-AD mice is consistent with the oxidative-stress-induced increase in SFPQ and TIA-1 levels in HeLa cells (Fig. 39), suggesting that at early stages of the disease pre-tangle pathology might induce these changes (Santacruz *et al.*, 2005).

During progression of the disease, SFPQ levels did not differ from controls at the early symptomatic phase of the disease. This study observed that, in 3xTg-AD mice at 12 mpi, there was a drastic reduction in SFPQ levels, comparable to the data from the postmortem human brain tissues from rpAD and sCJD patients. It was demonstrated that overexpression of SFPQ contributes to cell death, since it sensitizes neurons to neurotransmitter-mediated cell death (Xu *et al.*, 2005).

Reduction in TIA-1 (haplosufficiency) has been shown to be protective against neurodegeneration (Apicco *et al.*, 2018). The significant downregulation of TIA-1 observed at the terminal stage in 3xTg-AD mice and in spAD postmortem brains may exhibit a protective or compensatory effect (Fig. 39). The significant increase in TIA-1 in rpAD in comparison to spAD suggests higher neurodegeneration in rpAD cases.

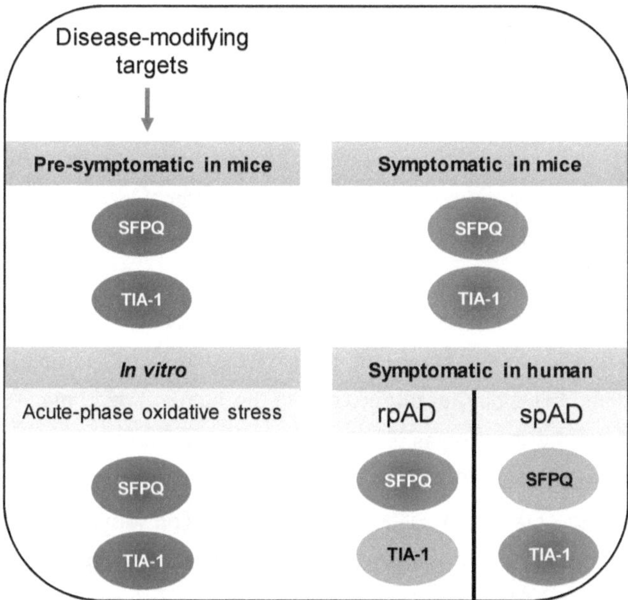

Figure 39: Comparative analysis of the differential expression of SFPQ and TIA-1 in mice and humans. The red boxes are indicating upregulated proteins, whereas the green boxes are representing downregulated proteins, with grey boxes indicating non-significant changes in relation to controls. Expression of SFPQ and TIA-1 was significantly elevated at the early pre-symptomatic phase (3 mpi) of the disease in 3xTg-AD mice, suggesting that these proteins could be of potential significance as early disease-modifying targets. The high expression levels of SFPQ and TIA-1 in mice were also observed in HeLa cells following exposure to oxidative stress. At symptomatic stage of the disease, differential regulation was observed for both SFPQ and TIA-1. The reduced SFPQ levels in the postmortem human brains of rpAD patients was also evident at late symptomatic stage in the 3xTg-AD mice.

In summary, the research findings from the current study confirmd the heterogeneity of AD entities. Proteomic analyses have identified several subtype-specific RBPs, but further studies are required to assess their clinical relevance.

4.5 Conclusion

RNA-binding proteins as key regulators in RNA processing and translational control may have pathophysiological functions in Alzheimer's disease. This study shows that the RNA-binding protein SFPQ (splicing factor proline and glutamine rich) is dysregulated at both the protein and mRNA level in the frontal cortex of patients dignosed with rpAD or sCJD as well as in the brains of 3xTg-AD mice. Co-immunofluorescence analysis in combination with confocal-laser scanning microscopy demonstrated nuclear depletion of SFPQ along with phospho-tau, particularly in cases of rapidly progressive AD. This nuclear depletion of both proteins was concomitantly associated with their cytoplasmic redistribution. Of note, association between SFPQ and tau in rpAD brain did not exclude the possible role of SFPQ in oligomerization and misfolding of tau protein. In the human brain, immunoreactivity of SFPQ co-localized with cytoplasmic TIA-1, which is a marker of stress granules. A similar translocation of SFPQ and phospho-tau into cytoplasmic TIA-1-positive stress granules was also obtained in cultured HeLa cells treated with sodium arsenite. Furthermore, the expression of human tau *in vitro* induced a significant reduction in SFPQ levels, suggesting a causal role of tau in downregulation of SFPQ. The findings from the current study indicate that dysregulated SFPQ in combination with pathological tau and aberrant dynamics of SGs represents an important pathway, which may contribute to the rapid progression of AD. The re-establishment of the normal localization of SFPQ may be a candidate for future therapeutic research strategies.

5 Annexure

Table 11: Patient details of Alzheimer's disease subtypes and non-demented controls

No.	Case	Age	Gender	Disease duration (y)	Braak stages	PMI (hr)
1	rpAD1	70	Male	<4	VI/ C	11:30
2	rpAD2	76	Female	<4	VI	18
3	rpAD3	79	Female	<4	V	05:30
4	rpAD4	83	Male	<4	VI/C	05:30
5	rpAD5	83	Male	<4	V/C	08:20
6	rpAD6	76	Male	<4	VI/C	06:30
7	rpAD7	77	Female	<4	IV/A	12
8	rpAD8	78	Male	<4	VI/C	03:30
9	spAD1	78	Male	>4	V/C	09:30
10	spAD2	72	Female	>4	V/C	09:30
11	spAD3	82	Female	>4	VI/B	01:45
12	spAD4	56	Female	>4	V/C	07
13	spAD5	87	Male	>4	V/C	07:05
14	spAD6	75	Female	>4	V/C	04:15
15	spAD7	93	Male	>4	V/C	03
16	spAD8	67	Female	>4	III/C	06:10
17	spAD9	90	Female	>4	IV/A	09:55
18	spAD10	83	Male	>4	III/0	07:25
19	Cont.1	69	Male	-	II/A	03:45
20	Cont.2	68	Male	-	I/0	10:55
21	Cont.3	64	Male	-	I/0	08:35
22	Cont.4	67	Male	-	I/0	14:40
23	Cont.5	74	Male	-	II/A	05:30
24	Cont.6	86	Male	-	II/A	05:30
25	Cont.7	73	Female	-	I/0	15:45
26	Cont.8	61	Male	-	I	04:30
27	Cont.9	77	Male	-	I/A	06:55

PMI (hr): Postmortem interval in hours, rpAD: rapidly progressive Alzheimer's disease, spAD: sporadic Alzheimer' disease.

Table 12: Details of sporadic Creutzfeldt Jakob disease subtype cases

No.	Case	Age	Gender	Disease duration (y)	Genotype	PMI (hr)
1	sCJD (MM1)1	65	Male	<1	MM/MV1	09:45
2	sCJD (MM1)2	74	Female	<1	MM/MV1	07:50
3	sCJD (MM1)3	61	Male	<1	MM/MV1	07
4	sCJD (MM1)4	66	Female	<1	MM/MV1	05:05
5	sCJD (MM1)5	74	Female	<1	MM/MV1	11
6	sCJD (MM1)6	74	Male	<1	MM/MV1	04:50
7	sCJD (VV2)1	66	Male	<1	VV2	15:30
8	sCJD (VV2)2	70	Female	<1	VV2	11
9	sCJD (VV2)3	72	Female	<1	VV2	06
10	sCJD (VV2)4	66	Female	<1	VV2	04

Table 13: Details of cases used for immunohistochemistry analysis

No.	Case	Age	Gender	Disease duration (y)	ABC score
1	rpAD1	59	Female	<4	A3, B3, C3
2	rpAD2	88	Female	<4	A3, B2, C3
3	rpAD3	76	Female	<4	A3, B3, C2
4	rpAD4	71	Female	<4	A2, B3, C3
5	rpAD5	84	Male	<4	A2, B3, C3
6	spAD1	87	Male	>4	A1, B3, C3
7	spAD2	77	Female	>4	A1, B3, C2
8	spAD3	69	Female	>4	A1, B2, C3
9	spAD4	62	Female	>4	A1, B1, C2
10	spAD5	85	Male	>4	A2, B3, C3
11	Cont.1	74	Female	-	-
12	Cont.2	87	Female	-	-
13	Cont.3	84	Female	-	-
14	Cont.4	82	Male	-	-
15	Cont.5	75	Male	-	-

Table 14: List of primer pairs used in the study

Gene name	Direction	Sequence
GAPDH	Forward	TGGGTGTGAACCATGAGAAGTA
GAPDH	Reverse	GAGTCCTTCCACGATACCAAAG
SFPQ	Forward	TGGGAAGTGACATGCGTACT
SFPQ	Reverse	TGTTTGGGCCTTCGTACTCT
TIA-1	Forward	AGTTTCCTGGCCTGCATTTC
TIA-1	Reverse	ACACTCGAGCTGTCTTTCCT
VCP	Forward	AAACGTATCCATGTGCTGCC
VCP	Reverse	ACTTTGAACTCCACAGCACG

5.1 RNA-binding protein candidates from mass spectrometry analysis.

Table 15: List of unique and common RNA-binding protein candidates identified in sporadic AD, rapidly progressive AD, and sporadic Creutzfeldt-Jakob disease

Disease Group	IDs	Uniprot Acc. No.	Protein names	Involvement in disease
spAD	AEDO	Q96SZ5	2-aminoethanethiol dioxygenase	
	AL1L1	O75891	Cytosolic 10-formyltetrahydrofolate dehydrogenase	
	AL4A1	P30038	Delta-1-pyrroline-5-carboxylate dehydrogenase, mitochondrial	Hyperprolinemia 2
	AP2M1	Q96CW1	AP-2 complex subunit mu	
	ARP2	P61160	Actin-related protein 2	
	ARRB1	P49407	Beta-arrestin-1	
	ATPD	P30049	ATP synthase subunit delta, mitochondrial	Mitochondrial complex V deficiency, nuclear type 5
	CAD13	P55290	Cadherin-13	

CADM4	Q8NFZ8	Cell adhesion molecule 4	
CANB1	P63098	Calcineurin subunit B type 1	
CAND1	Q86VP6	Cullin-associated NEDD8-dissociated protein 1	
CAPZB	P47756	F-actin-capping protein subunit beta	
CASA1	P47710	Alpha-S1-casein [Cleaved into: Casoxin-D]	
CAZA1	P52907	F-actin-capping protein subunit alpha-1	
CC50A	Q9NV96	Cell cycle control protein 50A	
CD47	Q08722	Leukocyte surface antigen CD47	
CLCA	P09496	Clathrin light chain A	
CLH2	P53675	Clathrin heavy chain 2	
CPLX2	Q6PUV4	Complexin-2	
CSN2	P61201	COP9 signalosome complex subunit 2	
CYFP2	Q96F07	Cytoplasmic FMR1-interacting protein 2	Epileptic encephalopathy, early infantile, 65
DBNL	Q9UJU6	Drebrin-like protein	
DC1L2	O43237	Cytoplasmic dynein 1 light intermediate chain 2	
DCXR	Q7Z4W1	L-xylulose reductase	Pentosuria
ECHA	P40939	Trifunctional enzyme subunit alpha, mitochondrial	Mitochondrial trifunctional protein deficiency
ENOB	P13929	Beta-enolase	Glycogen storage disease 13
FLNA	P21333	Filamin-A	Periventricular nodular heterotopia 1
FLOT1	O75955	Flotillin-1	
FUBP2	Q92945	Far upstream element-binding protein 2	
GANAB	Q14697	Neutral alpha-glucosidase AB	Polycystic kidney disease 3 with or without polycystic liver disease
GBB4	Q9HAV0	Guanine nucleotide-binding protein subunit beta-4	Charcot-Marie-Tooth disease, dominant, intermediate type, F
GBG3	P63215	Guanine nucleotide-binding protein G	
GDE	P35573	Glycogen debranching enzyme	Glycogen storage disease 3
GLU2B	P14314	Glucosidase 2 subunit beta	Polycystic liver disease 1 with or without kidney cysts
GNAQ	P50148	Guanine nucleotide-binding protein G	Capillary malformations, congenital
GPD1L	Q8N335	Glycerol-3-phosphate dehydrogenase 1-like protein	Brugada syndrome 2
GPDM	P43304	Glycerol-3-phosphate dehydrogenase, mitochondrial	
HINT1	P49773	Histidine triad nucleotide-binding protein 1	Neuromyotonia and axonal neuropathy, autosomal recessive
HS74L	O95757	Heat shock 70 kDa protein 4L	

HSP76	P17066	Heat shock 70 kDa protein 6	
ICAM5	Q9UMF0	Intercellular adhesion molecule 5	
IDH3B	O43837	Isocitrate dehydrogenase [NAD] subunit beta, mitochondrial	Retinitis pigmentosa 46
IF4A1	P60842	Eukaryotic initiation factor 4A-I	
IF4B	P23588	Eukaryotic translation initiation factor 4B	
IPYR	Q15181	Inorganic pyrophosphatase	
KCC2A	Q9UQM7	Calcium/calmodulin-dependent protein kinase type II subunit alpha	Mental retardation, autosomal dominant 53
KCRM	P06732	Creatine kinase M-type	
KT3K	Q9HA64	Ketosamine-3-kinase	
L1CAM	P32004	Neural cell adhesion molecule L1	Hydrocephalus due to stenosis of the aqueduct of Sylvius
LASP1	Q14847	LIM and SH3 domain protein 1	
LIGO1	Q96FE5	Leucine-rich repeat and immunoglobulin-like domain-containing nogo receptor-interacting protein 1	Mental retardation, autosomal recessive 64
MAOM	P23368	NAD-dependent malic enzyme, mitochondrial	
MAP4	P27816	Microtubule-associated protein 4	
MK03	P27361	Mitogen-activated protein kinase 3	
MT1F	P04733	Metallothionein-1F	
NCHL1	O00533	Neural cell adhesion molecule L1-like protein	
NCKP1	Q9Y2A7	Nck-associated protein 1	
NCKX2	Q9UI40	Sodium/potassium/calcium exchanger 2	
NDRG1	Q92597	Protein NDRG1	Charcot-Marie-Tooth disease 4D
NEDD8	Q15843	NEDD8	
NEGR1	Q7Z3B1	Neuronal growth regulator 1	
NP1L4	Q99733	Nucleosome assembly protein 1-like 4	
NTRI	Q9P121	Neurotrimin	
ODPA	P08559	Pyruvate dehydrogenase E1 component subunit alpha, somatic form, mitochondrial	Pyruvate dehydrogenase E1-alpha deficiency
OLA1	Q9NTK5	Obg-like ATPase 1	
OPA1	O60313	Dynamin-like 120 kDa protein, mitochondrial	Optic atrophy 1
OPCM	Q14982	Opioid-binding protein/cell adhesion molecule	Ovarian cancer
PAK1	Q13153	Serine/threonine-protein kinase PAK 1	Intellectual developmental disorder with macrocephaly, seizures, and speech delay
PAK3	O75914	Serine/threonine-protein kinase PAK 3	Mental retardation, X-linked 30
PCSK1	Q9UHG2	ProSAAS	
PCY2	Q99447	Ethanolamine-phosphate cytidylyltransferase	
PFKAL	P17858	ATP-dependent 6-phosphofructokinase, liver type	
PFKAP	Q01813	ATP-dependent 6-phosphofructokinase,	

			platelet type	
	PHIPL	Q96FC7	Phytanoyl-CoA hydroxylase-interacting protein-like	
	PLIN3	O60664	Perilipin-3	
	PPT1	P50897	Palmitoyl-protein thioesterase 1	Ceroid lipofuscinosis, neuronal, 1
	PTN11	Q06124	Tyrosine-protein phosphatase non-receptor type 11	LEOPARD syndrome 1
	QCR2	P22695	Cytochrome b-c1 complex subunit 2, mitochondrial	Mitochondrial complex III deficiency, nuclear 5
	RAB8B	Q92930	Ras-related protein Rab-8B	
	RALA	P11233	Ras-related protein Ral-A	
	RD23B	P54727	UV excision repair protein RAD23 homolog B	
	REEP5	Q00765	Receptor expression-enhancing protein 5	
	RHOC	P08134	Rho-related GTP-binding protein RhoC	
	SEMG1	P04279	Semenogelin-1	
	Septin-3	Q9UH03	Neuronal-specific septin-3	
	Septin-6	Q14141	Septin-6	
	Septin-9	Q9UHD8	Septin-9	
	SH3G1	Q99961	Endophilin-A2	
	SHLB2	Q9NR46	Endophilin-B2	
	SNAG	Q99747	Gamma-soluble NSF attachment protein	
	SPTN2	O15020	Spectrin beta chain, non-erythrocytic 2	Spinocerebellar ataxia 5
	SRC8	Q14247	Src substrate cortactin	
	SYNPO	Q8N3V7	Synaptopodin	
	TCAL5	Q5H9L2	Transcription elongation factor A protein-like 5	
	TCPB	P78371	T-complex protein 1 subunit beta	
	TCPQ	P50990	T-complex protein 1 subunit theta	
	TCTP	P13693	Translationally-controlled tumor protein	
rpAD	4F2	P08195	4F2 cell-surface antigen heavy chain	
	ACTA	P62736	Actin, aortic smooth muscle	
	ACY2	P45381	Aspartoacylase	Canavan disease
	ADDA	P35611	Alpha-adducin	
	ADT2	P05141	ADP/ATP translocase 2	
	AL7A1	P49419	Alpha-aminoadipic semialdehyde dehydrogenase	Pyridoxine-dependent epilepsy
	AMER2	Q8N7J2	APC membrane recruitment protein 2	
	AMPL	P28838	Cytosol aminopeptidase	
	ANK2	Q01484	Ankyrin-2	Long QT syndrome 4
	AOFA	P21397	Amine oxidase [flavin-containing] A	Brunner syndrome
	AOFB	P27338	Amine oxidase [flavin-containing] B	
	AP180	O60641	Clathrin coat assembly protein AP180	
	ASAH1	Q13510	Acid ceramidase	Farber lipogranulomatosis
	AT2A2	P16615	Sarcoplasmic/endoplasmic reticulum calcium ATPase 2	Acrokeratosis verruciformis
	ATP5H	O75947	ATP synthase subunit d, mitochondrial	

ATP5J	P18859	ATP synthase-coupling factor 6, mitochondrial	
ATP5L	O75964	ATP synthase subunit g, mitochondrial	
CALX	P27824	Calnexin	
CD44	P16070	CD44 antigen	
CDS2	O95674	Phosphatidate cytidylyltransferase 2	
CLD11	O75508	Claudin-11	
CMC1	O75746	Calcium-binding mitochondrial carrier protein Aralar1	Epileptic encephalopathy, early infantile, 39
CO4A	P0C0L4	Complement C4-A	Complement component 4A deficiency
CO4B	P0C0L5	Complement C4-B	Systemic lupus erythematosus
COX41	P13073	Cytochrome c oxidase subunit 4 isoform 1, mitochondrial	
CPNS1	P04632	Calpain small subunit 1	
CUTA	O60888	Protein CutA	
DC1I2	Q13409	Cytoplasmic dynein 1 intermediate chain 2	
DDTL	A6NHG4	D-dopachrome decarboxylase-like protein	
EF2	P13639	Elongation factor 2	Spinocerebellar ataxia 26
ERMIN	Q8TAM6	Ermin	
FIS1	Q9Y3D6	Mitochondrial fission 1 protein	
GBG2	P59768	Guanine nucleotide-binding protein G	
GHC1	Q9H936	Mitochondrial glutamate carrier 1	Epileptic encephalopathy, early infantile, 3
GPM6B	Q13491	Neuronal membrane glycoprotein M6-b	
HEBP1	Q9NRV9	Heme-binding protein 1	
HNRPD	Q14103	Heterogeneous nuclear ribonucleoprotein D0	
HYEP	P07099	Epoxide hydrolase 1	
IMB1	Q14974	Importin subunit beta-1	
LAMP1	P11279	Lysosome-associated membrane glycoprotein 1	
LANC2	Q9NS86	LanC-like protein 2	
MAG	P20916	Myelin-associated glycoprotein	Spastic paraplegia 75, autosomal recessive
MRP	P49006	MARCKS-related protein	
MTAP2	P11137	Microtubule-associated protein 2	
MTPN	P58546	Myotrophin	
NDKA	P15531	Nucleoside diphosphate kinase A	
NDUA4	O00483	Cytochrome c oxidase subunit NDUFA4	Leigh syndrome
ODO2	P36957	Dihydrolipoyllysine-residue succinyltransferase component of 2-oxoglutarate dehydrogenase complex, mitochondrial	Platelet-activating factor acetylhydrolase IB subunit beta
PA1B2	P68402		
PCBP2	Q15366	Poly(rC)-binding protein 2	
PGAM2	P15259	Phosphoglycerate mutase 2	Glycogen storage disease 10

	Gene	Accession	Protein	Disease
	PHB	P35232	Prohibitin	
	PHB2	Q99623	Prohibitin-2	
	PI42A	P48426	Phosphatidylinositol 5-phosphate 4-kinase type-2 alpha	
	PRIO	P04156	Major prion protein	prion diseases, like: Creutzfeldt-Jakob disease
	QCR1	P31930	Cytochrome b-c1 complex subunit 1, mitochondrial	
	QCR6	P07919	Cytochrome b-c1 complex subunit 6, mitochondrial	
	RAP1A	P62834	Ras-related protein Rap-1A	
	SCG1	P05060	Secretogranin-1	
	SCG2	P13521	Secretogranin-2	
	SDHA	P31040	Succinate dehydrogenase [ubiquinone] flavoprotein subunit, mitochondrial	Mitochondrial complex II deficiency
	Septin-4	O43236	Septin-4	
	SERA	O43175	D-3-phosphoglycerate dehydrogenase	Phosphoglycerate dehydrogenase deficiency
	SIRB1	O00241	Signal-regulatory protein beta-1	
	SNG3	O43761	Synaptogyrin-3	
	SV2A	Q7L0J3	Synaptic vesicle glycoprotein 2A	
	SV2B	Q7L1I2	Synaptic vesicle glycoprotein 2B	
	SYNJ1	O43426	Synaptojanin-1	Parkinson disease 20, early-onset
	TBA1B	P68363	Tubulin alpha-1B chain	
	TCPE	P48643	T-complex protein 1 subunit epsilon	Neuropathy, hereditary sensory, with spastic paraplegia, autosomal recessive
	TENA	P24821	Tenascin	Deafness, autosomal dominant, 56
	VA0D1	P61421	V-type proton ATPase subunit d 1	
	VAT1	Q99536	Synaptic vesicle membrane protein VAT-1 homolog Voltage-dependent anion-selective channel protein 1	Voltage-dependent anion-selective channel protein 2 Voltage-dependent anion-selective channel protein 3
	VDAC1	P21796		
	VDAC2	P45880		
	VDAC3	Q9Y277		
sCJD	**A2MG**	P01023	Alpha-2-macroglobulin	
	AACT	P01011	Alpha-1-antichymotrypsin	
	ACO13	Q9NPJ3	Acyl-coenzyme A thioesterase 13	
	ADHX	P11766	Alcohol dehydrogenase class-3	
	AL1A1	P00352	Retinal dehydrogenase 1	
	ALDR	P15121	Aldo-keto reductase family 1 member B1	
	APEX1	P27695	DNA-(apurinic or apyrimidinic site) lyase	
	APOA1	P02647	Apolipoprotein A-I	High density lipoprotein deficiency 2

ARP3	P61158	Actin-related protein 3	
ASGL1	Q7L266	Isoaspartyl peptidase/L-asparaginase	
BlEA	P53004	Biliverdin reductase A	Hyperbiliverdinemia
CAB39	Q9Y376	Calcium-binding protein 39	
CATA	P04040	Catalase	Acatalasemia
CD59	P13987	CD59 glycoprotein	Hemolytic anemia, CD59-mediated, with or without poly-neuropathy
CLIC4	Q9Y696	Chloride intracellular channel protein 4	
CMBL	Q96DG6	Carboxymethylenebutenolidase homolog	
CPPED	Q9BRF8	Serine/threonine-protein phosphatase CPPED1	
DOPD	P30046	D-dopachrome decarboxylase	
DYL2	Q96FJ2	Dynein light chain 2, cytoplasmic	
EFHD2	Q96C19	EF-hand domain-containing protein D2	
FABP7	O15540	Fatty acid-binding protein, brain	
FAHD1	Q6P587	Acylpyruvase FAHD1, mitochondrial	
FBX2	Q9UK22	F-box only protein 2	
FIBB	P02675	Fibrinogen beta chain	Congenital afibrinogenemia
FIBG	P02679	Fibrinogen gamma chain	Congenital afibrinogenemia
FKBP4	Q02790	Peptidyl-prolyl cis-trans isomerase FKBP4	
G6PD	P11413	Glucose-6-phosphate 1-dehydrogenase	Anemia, non-spherocytic hemolyt-ic, due to G6PD deficiency
GBRL2	P60520	Gamma-aminobutyric acid receptor-associated protein-like 2	
GGCT	O75223	Gamma-glutamylcyclotransferase	
GNAI2	P04899	Guanine nucleotide-binding protein G	
GNPI1	P46926	Glucosamine-6-phosphate isomerase 1	
GSHB	P48637	Glutathione synthetase	Glutathione synthetase deficiency
HMGB1	P09429	High mobility group protein B1	
HNRPK	P61978	Heterogeneous nuclear ribonucleoprotein K	Au-Kline syndrome
IDHC	O75874	Isocitrate dehydrogenase [NADP] cytoplasmic	Glioma
IDHP	P48735	Isocitrate dehydrogenase [NADP], mitochondrial	D-2-hydroxyglutaric aciduria 2
IGLL5	B9A064	Immunoglobulin lambda-like polypeptide 5	
ILF2	Q12905	Interleukin enhancer-binding factor 2	
KAD3	Q9UIJ7	GTP: AMP phosphotransferase AK3, mitochondrial	Phospholysine phosphohistidine inorganic pyrophos-phate phosphatase
LHPP	Q9H008		
LKHA4	P09960	Leukotriene A-4 hydrolase	
MPI	P34949	Mannose-6-phosphate isomerase	Congenital disorder of glycosylation 1B

MT2	P02795	Metallothionein-2	
NNRE	Q8NCW5	NAD(P)H-hydrate epimerase	Encephalopathy, progressive, early-onset, with brain edema and/or leu-koencephalopathy
NQO2	P16083	Ribosyldihydronicotinamide dehydrogenase	
NUDT5	Q9UKK9	ADP-sugar pyrophosphatase	
PDCD6	O75340	Programmed cell death protein 6	
PITH1	Q9GZP4	PITH domain-containing protein 1	
PROF2	P35080	Profilin-2	
PSA1	P25786	Proteasome subunit alpha type-1	
PSA2	P25787	Proteasome subunit alpha type-2	
PSA4	P25789	Proteasome subunit alpha type-4	
PSA5	P28066	Proteasome subunit alpha type-5	
PSA6	P60900	Proteasome subunit alpha type-6	
PSA7	O14818	Proteasome subunit alpha type-7	
PSB1	P20618	Proteasome subunit beta type-1	
PSB3	P49720	Proteasome subunit beta type-3	
PSB5	P28074	Proteasome subunit beta type-5	
PTGR1	Q14914	Prostaglandin reductase 1	
RAB21	Q9UL25	Ras-related protein Rab-21	
RAB5B	P61020	Ras-related protein Rab-5B	
RASK	P01116	GTPase KRas	Leukemia, acute myelogenous
RB11B	Q15907	Ras-related protein Rab-11B	Neurodevelopmen-tal disorder with ataxic gait, absent speech, and de-creased cortical white matter
RHOA	P61586	Transforming protein RhoA	
SH3L2	Q9UJC5	SH3 domain-binding glutamic acid-rich-like protein 2	
SKP1	P63208	S-phase kinase-associated protein 1	
SPB6	P35237	Serpin B6	Deafness, autosomal recessive, 91
SYWC	P23381	Tryptophan--tRNA ligase, cytoplasmic	Neuronopathy, distal hereditary motor, 9
TOLIP	Q9H0E2	Toll-interacting protein	
UBC12	P61081	NEDD8-conjugating enzyme Ubc12	
rpAD and sCJD	**Common between rpAD and SCJD**		
A1AG1	P02763	Alpha-1-acid glycoprotein 1	
A1AT	P01009	Alpha-1-antitrypsin	Alpha-1-antitrypsin deficiency
ACTN1	P12814	Alpha-actinin-1	Bleeding disorder, platelet-type 15
ACYP2	P14621	Acylphosphatase-2	
AK1A1	P14550	Aldo-keto reductase family 1 member A1	

125

CD81	P60033	CD81 antigen	Immunodeficiency, common variable, 6
CYTB	P04080	Cystatin-B	Epilepsy, progressive myoclonic 1
DNJC5	Q9H3Z4	DnaJ homolog subfamily C member 5	Ceroid lipofuscinosis, neuronal, 4B
FRIL	P02792	Ferritin light chain	Hyperferritinemia with or without cataract
FSCN1	Q16658	Fascin	
GLTP	Q9NZD2	Glycolipid transfer protein	
GSTM2	P28161	Glutathione S-transferase Mu 2	
HPLN1	P10915	Hyaluronan and proteoglycan link protein 1	
LIS1	P43034	Platelet-activating factor acetylhydrolase IB subunit alpha	Lissencephaly 1
OTUB1	Q96FW1	Ubiquitin thioesterase OTUB1	
PDIA3	P30101	Protein disulfide-isomerase A3	
PEA15	Q15121	Astrocytic phosphoprotein PEA-15	
PIPNA	Q00169	Phosphatidylinositol transfer protein alpha isoform	
PPIA	P62937	Peptidyl-prolyl cis-trans isomerase A	
PTGDS	P41222	Prostaglandin-H2 D-isomerase	
QOR	Q08257	Quinone oxidoreductase	
RAB5C	P51148	Ras-related protein Rab-5C	
RIDA	P52758	2-iminobutanoate/2-iminopropanoate deaminase	
SCRN1	Q12765	Secernin-1	
SERC	Q9Y617	Phosphoserine aminotransferase	Phosphoserine aminotransferase deficiency
SFPQ	P23246	Splicing factor, proline- and glutamine-rich	
SH3L1	O75368	SH3 domain-binding glutamic acid-rich-like protein	
SNAA	P54920	Alpha-soluble NSF attachment protein	
SNAB	Q9H115	Beta-soluble NSF attachment protein	
SODM	P04179	Superoxide dismutase [Mn], mitochondrial	Microvascular complications of diabetes 6
TKT	P29401	Transketolase	Short stature, developmental delay, and congenital heart defects
TRFE	P02787	Serotransferrin	
UGPA	Q16851	UTP--glucose-1-phosphate uridylyltransferase	
VATH	Q9UI12	V-type proton ATPase subunit H	
VCP	P55072	Valosin containing protein	Inclusion body myopathy with early-onset Paget disease with or without frontotemporal dementia 1

spAD and sCJD	**Common between spAD and sCJD**			
		O43488	Aflatoxin B1 aldehyde reductase member 2	
	CYBP	Q9HB71	Calcyclin-binding protein	
	DDAH2	O95865	N(G), N(G)-dimethylarginine dimethyla-minohydrolase 2	
	ECH1	Q13011	Delta(3,5)-Delta (2,4)-dienoyl-CoA isomeras, mitochondrial	
	ENOPH	Q9UHY7	Enolase-phosphatase E1	
	FPPS	P14324	Farnesyl pyrophosphate synthase	Porokeratosis 9, multiple types (POROK9)
	GLO2	Q16775	Hydroxyacylglutathione hydrolase, mitochondrial	
	GLOD4	Q9HC38	Glyoxalase domain-containing protein 4	
	GMFB	P60983	Glia maturation factor beta	
	GSTM1	P09488	Glutathione S-transferase Mu 1	
	GSTM3	P21266	Glutathione S-transferase Mu 3	
	LGUL	Q04760	Lactoylglutathione lyase	
	LSAMP	Q13449	Limbic system-associated membrane protein	
	MK01	P28482	Mitogen-activated protein kinase 1	
	NCDN	Q9UBB6	Neurochondrin	
	PGM2L	Q6PCE3	Glucose 1,6-bisphosphate synthase	
	PP2BB	P16298	Serine/threonine-protein phosphatase 2B catalytic subunit beta isoform	
	RAB2A	P61019	Ras-related protein Rab-2A	
	RAB5A	P20339	Ras-related protein Rab-5A	
	TAGL	Q01995	Transgelin	
spAD and rpAD	**Common between spAD and rpAD**			
	AKA12	Q02952	A-kinase anchor protein 12	
	ANXA7	P20073	Annexin A7	
	APOE	P02649	Apolipoprotein E	Hyperlipoproteinemia 3
	AQP4	P55087	Aquaporin-4	
	AT2B1	P20020	Plasma membrane calcium-transporting ATPase 1	
	AT2B2	Q01814	Plasma membrane calcium-transporting ATPase 2	
	AT2B3	Q16720	Plasma membrane calcium-transporting ATPase 3	Spinocerebellar ataxia, X-linked 1
	AT2B4	P23634	Plasma membrane calcium-transporting ATPase 4	
	AT5F1	P24539	ATP synthase F	
	ATPG	P36542	ATP synthase subunit gamma, mitochondrial	
	BCAS1	O75363	Breast carcinoma-amplified sequence 1	
	BSN	Q9UPA5	Protein bassoon	
	CADM2	Q8N3J6	Cell adhesion molecule 2	
	CXA1	P17302	Gap junction alpha-1 protein	Oculodentodigital dysplasia
	EAA1	P43003	Excitatory amino acid transporter 1	Episodic ataxia 6
	EAA2	P43004	Excitatory amino acid transporter 2	Epileptic encephalopathy,

			early infantile, 41
F10A1	P50502	Hsc70-interacting protein	
FIBA	P02671	Fibrinogen alpha chain	Congenital afibrinogenemia
HECAM	Q14CZ8	Hepatocyte cell adhesion molecule	Leukoencephalopathy, megalencephalic, with subcortical cysts, 2A
MAON	Q16798	NADP-dependent malic enzyme, mitochondrial	
MOG	Q16653	Myelin-oligodendrocyte glycoprotein	Narcolepsy 7
MT1G	P13640	Metallothionein-1G	
MT3	P25713	Metallothionein-3	
NCAM2	O15394	Neural cell adhesion molecule 2	
NCAN	O14594	Neurocan core protein	
NHRF1	O14745	Na (+)/H (+) exchange regulatory cofactor NHE-RF1	Nephrolithiasis/osteoporosis, hypophosphatemic, 2
NPTN	Q9Y639	Neuroplastin	
NRCAM	Q92823	Neuronal cell adhesion molecule	
OXR1	Q8N573	Oxidation resistance protein 1	
PADI2	Q9Y2J8	Protein-arginine deiminase type-2	
PALM	O75781	Paralemmin-1	
PDIA6	Q15084	Protein disulfide-isomerase A6	
PFKAM	P08237	ATP-dependent 6-phosphofructokinase, muscle type	Glycogen storage disease 7
PLCB1	Q9NQ66	1-phosphatidylinositol 4,5-bisphosphate phosphodiesterase beta-1	Epileptic encephalopathy, early infantile, 12
PRRT2	Q7Z6L0	Proline-rich transmembrane protein 2	Episodic kinesigenic dyskinesia 1
PTMA	P06454	Prothymosin alpha	
PTPRZ	P23471	Receptor-type tyrosine-protein phosphatase zeta	
RAB6B	Q9NRW1	Ras-related protein Rab-6B	
RTN1	Q16799	Reticulon-1	
RTN4	Q9NQC3	Reticulon-4	
S10AD	Q99584	Protein S100-A13	
SH3L3	Q9H299	SH3 domain-binding glutamic acid-rich-like protein 3	
SNG1	O43759	Synaptogyrin-1	
TENR	Q92752	Tenascin-R	
TYB4	P62328	Thymosin beta-4	
VPP1	Q93050	V-type proton ATPase 116 kDa subunit a isoform 1	

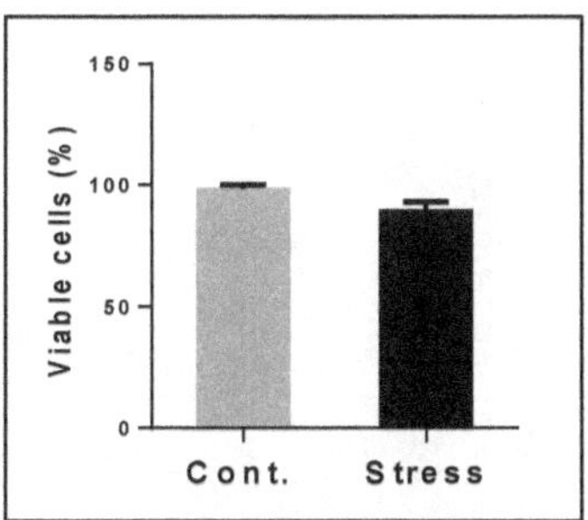

Figure 40: Cell viability assay: Trypan blue exclusion test was used to estimate the cell viability. The percentage of live cells was calculated in control (untreated) and stress cells (treated with 0.6 mM NaAso$_2$ for 60 min).

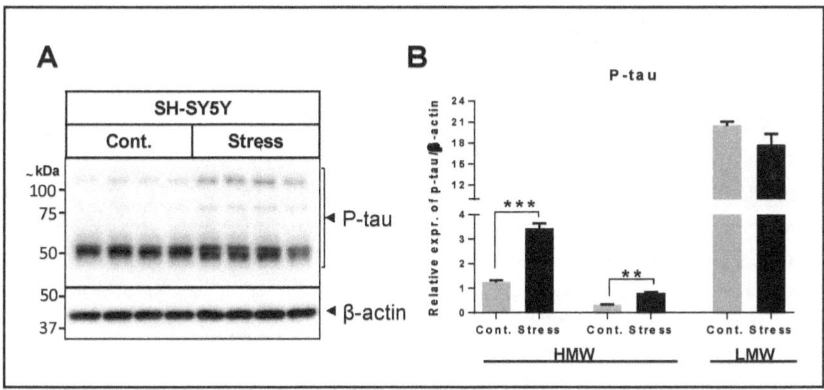

Figure 41: Stress induced increase in tau phosphorylation in SH-SY5Y cells. A) Representative immunoblots of phospho-tau in control (untreated) and stress (arsenite treated) cells. The cells were plated in 6-well plates (2x10^5) for 24 hrs and lysed in cell-lysis buffer supplemented with protease and phosphatase inhibitors, and levels of phospho-tau were analysed by immunoblotting analysis. Intensity levels were normalized to β-actin. **B)** The densitometry was performed using Image Lab software. A significant increase was observed in the levels of phospho-tau in high molecular weight range (HMW = 65-250 kDa). No significant changes were observed for low molecular weight range (LMW < 65 kDa). Statistical tests (unpaired t-test) were applied in GraphPad prism, **p < 0.01, ***p < 0.001.

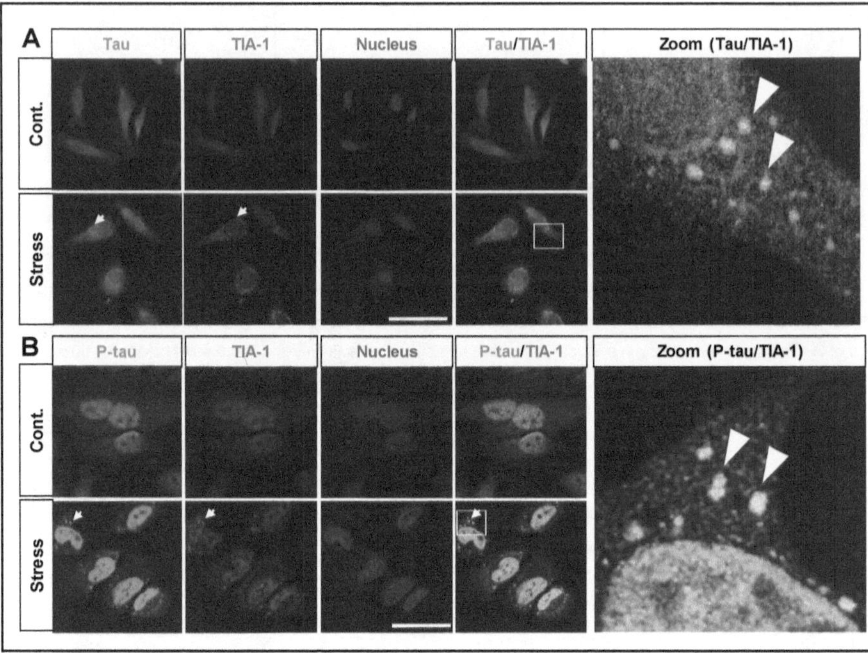

Figure 42: Tau and phospho-tau are recruited into SGs. A and B) Stress was induced with sodium arsenite and cells were co-immunostained with primary antibodies specific for total tau, phospho-tau and TIA-1, followed by incubation with AlexaFlour 488 and AlexaFlour 546 secondary antibodies. High magnification showing the expression of tau/TIA-1 and p-tau/TIA-1 for closer details in stress induced cells. Examples of SGs are indicated by the *arrows*, scale bar = 25 µm for tau and 10 µm for p-tau.

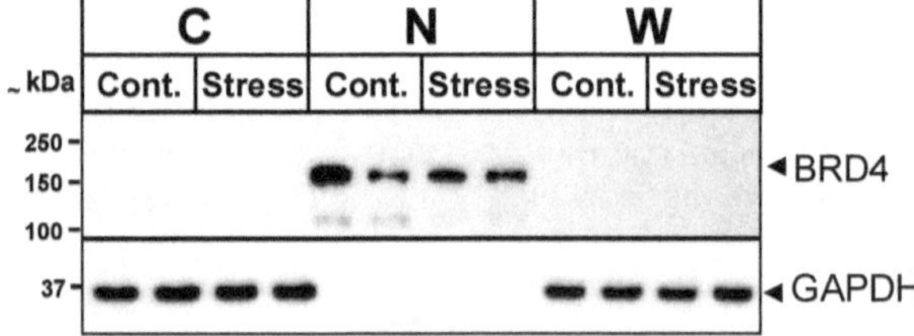

Figure 43: Subcellular fractionation after stress induction: **A)** Representative immunoblots for nuclear (BRD4) and cytoplasmic (GAPDH) markers in control (untreated) and stress (arsenite treated) cells, after subcellular fractionation by REAP method. Isolated fractions were abbreviated as C: cytoplasmic extract, N: nuclear extract, and W: whole cell extract. There was no cross contamination observed between nuclear and cytoplasmic markers.

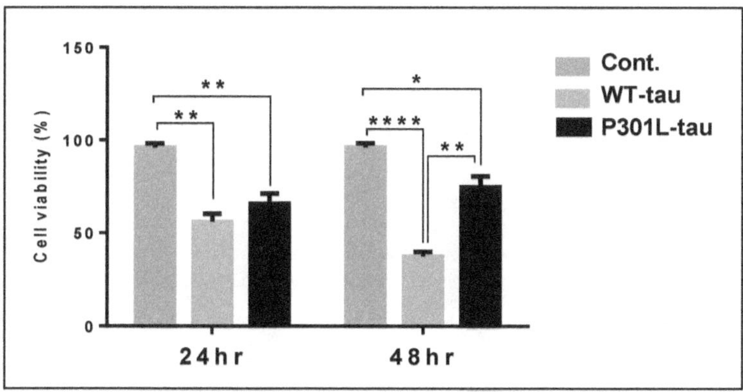

Figure 44: Cell Viability assay: The cell viability was estimated by MTS assay after expression of human-tau (both WT-tau and P301L-tau) in comparison to control at 24- and 48 hrs post-transfection. One-way ANOVA followed by Tukey post-hoc analysis was used, $*p < 0.05$, $**p < 0.01$, $***p < 0.001$, $****p < 0.0001$.

Table 16: Tau-up regulated proteins: list of unique and common proteins (with their -log$_{10}$p-values), that were upregulated after expression of either WT-tau or P301L-tau

No.	UniProt ID	UniProt Accession	Protein names	(WT vs Cont.) (p-values)	(P301L vs Cont.) (p-values)
1	IST1	P53990	IST1 homolog	2.80404	
2	XCT	Q9UPY5	Cystine/glutamate transporter	2.70942	
3	ITPR3	Q14573	Inositol 1,4,5-trisphosphate receptor type 3	3.74147	
4	CAD13	P55290	Cadherin-13	3.29636	
5	TMEDA	P49755	Transmembrane emp24 domain-containing protein 10	2.79315	
6	APC7	Q9UJX3	Anaphase-promoting complex subunit 7	2.64649	
7	PXDC2	Q6UX71	Plexin domain-containing protein 2	2.60679	
8	PLCA	Q99943	1-acyl-sn-glycerol-3-phosphate acyltransferase alpha	2.77202	
9	NUDC1	Q96RS6-2	NudC domain-containing protein 1	2.58966	
10	SYUG	O76070	Gamma-synuclein	3.30288	
11	A0A1W2PS4 3	A0A1W2PS 43	Lysosome membrane protein 2	3.00432	
12	F8VX04	F8VX04	Sodium-coupled neutral amino acid transporter 1	3.15582	
13	RAB18	Q9NP72	Ras-related protein Rab-18	2.60603	
14	Q5VZR0	Q5VZR0	Golgi-associated plant patho-genesis-related protein 1	2.6549	
15	CD44	P16070	CD44 antigen	2.89138	
16	COX2	P00403	Cytochrome c oxidase subunit 2	2.89006	
17	PYRG1	P17812	CTP synthase 1	2.64826	
18	C9JYN0	C9JYN0	Synaptophysin-like protein 1	3.73999	
19	E7ER44	E7ER44	Lactotransferrin		5.61449

131

20	**K7ENL2**	K7ENL2	WW domain-binding protein 2		3.25965
21	**IF16**	Q16666	Gamma-interferon-inducible protein 16		2.54071
22	**A0A1B0GW C0**	A0A1B0G WC0	Carnitine O-palmitoyltransferase 2, mitochondrial		3.83877
23	**A0A087WX9 7**	A0A087WX 97	Bcl-2-like protein 13		3.19665
24	**RND3**	P61587	Rho-related GTP-binding protein RhoE		3.65369
25	**PODXL**	O00592	Podocalyxin		5.52997
26	**MOT1**	P53985	Monocarboxylate transporter 1		4.65667
27	**CLIC4**	Q9Y696	Chloride intracellular channel protein 4		4.5989
28	**A0A075B73 0**	A0A075B7 30	Epiplakin		2.80249
29	**S38A2**	Q96QD8	Sodium-coupled neutral amino acid transporter 2		3.7117
30	**TPBG**	Q13641	Trophoblast glycoprotein		3.97711
31	**LAT1**	Q01650	Large neutral amino acids transporter small subunit 1		6.21328
32	**SQOR**	Q9Y6N5	Sulfide: quinone oxidoreductase, mitochondrial		3.5484
33	**E9PEB5**	E9PEB5	Far upstream element-binding protein 1		3.65134
34	**AAAT**	Q15758	Neutral amino acid transporter B		4.73081
35	**AHNK2**	Q8IVF2	Protein AHNAK2		2.93411
36	**RTN4**	Q9NQC3-2	Reticulon-4		2.78976
37	**AT1A1**	P05023-4	Sodium/potassium-transporting ATPase subunit alpha-1		2.80245
38	**A0A087X054**	A0A087X0 54	Hypoxia up-regulated protein 1		2.67896
39	**J3KPF3**	J3KPF3	4F2 cell-surface antigen heavy chain		3.40599
40	**VINC**	P18206	Vinculin		2.67223
41	**SYWC**	P23381-2	Tryptophan--tRNA ligase, cytoplasmic		2.87585
42	**UAP1**	Q16222	UDP-N-acetylhexosamine pyrophosphorylase		2.53642
43	**PLST**	P13797	Plastin-3		2.62095
Common in both WT- and P301L-tau expressing cells					
44	**Tau**	P10636-6	Microtubule-associated protein tau	9.66209	9.99752
45	**STX12**	Q86Y82	Syntaxin-12	4.38935	3.90083
46	**A0A0A0MRJ 7**	A0A0A0MR J7	Coagulation factor V	2.79996	2.55125
47	**TSP1**	P07996	Thrombospondin-1	4.63034	3.69623
48	**STOM**	P27105	Erythrocyte band 7 integral membrane protein	2.63921	4.70864
49	**AT1B1**	P05026	Sodium/potassium-transporting ATPase subunit beta-1	4.36966	4.04487
50	**SNG2**	O43760	Synaptogyrin-2	3.38169	3.37281
51	**SODM**	P04179	Superoxide dismutase [Mn], mitochondrial	4.35392	4.21758
52	**E9PR17**	E9PR17	CD59 glycoprotein	4.06777	5.10992
53	**MOT4**	O15427	Monocarboxylate transporter 4	3.57067	3.15312
54	**RAI3**	Q8NFJ5	Retinoic acid-induced protein 3	5.11684	3.4051

55	RTN3	O95197-3	Reticulon-3	2.89615	2.83528
56	FLNB	O75369-8	Filamin-B	4.65931	4.00561
57	QCR1	P31930	Cytochrome b-c1 complex subunit 1, mitochondrial	3.85388	5.56346
58	B4DKB2	B4DKB2	Endothelin-converting enzyme 1	2.89353	2.60173
59	VDAC2	P45880	Voltage-dependent anion-selective channel protein 2	3.02634	3.84069
60	ANXA3	P12429	Annexin A3	2.89038	4.15502
61	BIP	P11021	Endoplasmic reticulum chaperone BiP	4.09375	5.35755
62	ANXA4	P09525	Annexin A4	4.25635	3.6306
63	A0A0G2JIW1	A0A0G2JIW1	Heat shock 70 kDa protein 1B	3.97859	2.54103

Table 17: Tau down-regulated proteins: list of unique and common proteins (with their -log$_{10}$p-values), that were down-regulated after expression of either WT-tau or P301L-tau

No.	UniProt ID	UniProt Accession	Protein names	WT vs Cont. (p-values)	P301L vs Cont. (p-values)
1	CAPZB	P47756-2	F-actin-capping protein subunit beta	2.90632	
2	CLIC1	O00299	Chloride intracellular channel protein 1	2.74343	
3	1433E	P62258	14-3-3 protein epsilon	2.97992	
4	PDLI7	Q9NR12	PDZ and LIM domain protein 7	2.87187	
5	DHPR	P09417	Dihydropteridine reductase	3.29976	
6	TBB4B	P68371	Tubulin beta-4B chain	2.61161	
7	CYBP	Q9HB71	Calcyclin-binding protein	4.20659	
8	GSH0	P48507	Glutamate--cysteine ligase regulatory subunit	3.64493	
9	IDHC	O75874	Isocitrate dehydrogenase [NADP] cytoplasmic	3.18656	
10	CRK	P46108	Adapter molecule crk	2.66295	
11	LKHA4	P09960	Leukotriene A-4 hydrolase	3.07686	
12	LDHA	P00338	L-lactate dehydrogenase A chain	2.60496	
13	COF1	P23528	Cofilin-1	2.86394	
14	PSME3	P61289	Proteasome activator complex subunit 3	2.59093	
15	XPO6	Q96QU8	Exportin-6	2.63751	
16	PFKAM	P08237	ATP-dependent 6-phosphofructokinase, muscle type	2.77411	
17	METK2	P31153	S-adenosylmethionine synthase isoform type-2	2.59551	
18	I3L0H8	I3L0H8	ATP-dependent RNA helicase DDX19A	3.00712	
19	A0A087WYT3	A0A087WYT3	Prostaglandin E synthase 3	6.25017	
20	MBB1A	Q9BQG0	Myb-binding protein 1A	3.31109	
21	GANAB	Q14697-2	Neutral alpha-glucosidase AB	3.99331	
22	RPAC1	O15160	DNA-directed RNA polymerases I and III subunit RPAC1	2.95675	
23	6PGD	P52209	6-phosphogluconate	3.15286	

			dehydrogenase, decarboxylating	
24	**F2Z2Y4**	F2Z2Y4	Pyridoxal kinase	3.349
25	**KTHY**	P23919	Thymidylate kinase	4.05072
26	**ACADM**	P11310-2	Medium-chain specific acyl-CoA dehydrogenase, mito-chondrial	3.37726
27	**RT27**	Q92552	28S ribosomal protein S27, mitochondrial	2.65651
28	**A0A1B0GW77**	A0A1B0GW77	Alpha-aminoadipic semialdehyde dehydrogenase	2.57993
29	**IPYR**	Q15181	Inorganic pyrophosphatase	2.56092
30	**KPYM**	P14618	Pyruvate kinase PKM	3.3471
31	**TWF1**	Q12792	Twinfilin-1	3.16219
32	**MCM6**	Q14566	DNA replication licensing fac-tor MCM6	2.98343
33	**I3L2B0**	I3L2B0	Clustered mitochondria protein homolog	3.14353
34	**KPRA**	Q14558	Phosphoribosyl pyrophos-phate synthase-associated protein 1	2.88641
35	**ERO1A**	Q96HE7	ERO1-like protein alpha	3.48398
36	**AP3D1**	O14617	AP-3 complex subunit delta-1	2.64148
37	**Q5QPM7**	Q5QPM7	Proteasome inhibitor PI31 subunit	2.779
38	**TNPO3**	Q9Y5L0	Transportin-3	3.19029
39	**NAT10**	Q9H0A0	RNA cytidine acetyltransferase	2.60558
40	**F6WQW2**	F6WQW2	Ran-specific GTPase-activating protein	2.6097
41	**HAT1**	O14929	Histone acetyltransferase type B catalytic subunit	2.61858
42	**E7ESZ7**	E7ESZ7	NADH dehydrogenase [ubiq-uinone] 1 alpha subcomplex subunit 10, mitochondrial	3.34015
43	**A0A1B0GWE8**	A0A1B0GWE8	Cathepsin D	3.59333
44	**APEX1**	P27695	DNA-(apurinic or apyrimidinic site) lyase	2.573
45	**LNP**	Q9C0E8	Endoplasmic reticulum junc-tion formation protein lunapark	3.024
46	**PARK7**	Q99497	Protein/nucleic acid deglycase DJ-1	3.53845
47	**PP2AA**	P67775	Serine/threonine-protein phosphatase 2A catalytic sub-unit alpha isoform	2.85783
48	**DUS23**	Q9BVJ7	Dual specificity protein phosphatase 23	2.6001
49	**NUP43**	Q8NFH3	Nucleoporin Nup43	2.92109
50	**TF3C4**	Q9UKN8	General transcription factor 3C polypeptide 4	2.85149
51	**UBA3**	Q8TBC4	NEDD8-activating enzyme E1 catalytic subunit	2.85788
52	**B1AH49**	B1AH49	3-mercaptopyruvate sulfurtransferase	4.16312
53	**ABCD3**	P28288	ATP-binding cassette sub-family D member 3	2.89151
54	**Q5QPR3**	Q5QPR3	Cyclin-dependent kinase 11A	3.07425

55	**D6RG13**	D6RG13	40S ribosomal protein S3a	3.40351	
56	**B7Z4B8**	B7Z4B8	Heterogeneous nuclear ribo-nucleoprotein U-like protein 1	3.80247	
57	**AN32E**	Q9BTT0	Acidic leucine-rich nuclear phosphoprotein 32 family member E	3.0431	
58	**HPBP1**	Q9NZL4	Hsp70-binding protein 1	3.14118	
59	**PAIRB**	Q8NC51	Plasminogen activator inhibitor 1 RNA-binding protein	2.72168	
60	**CDK1**	P06493	Cyclin-dependent kinase 1	6.03772	
61	**X6RLT1**	X6RLT1	Negative elongation factor C/D	2.93622	
62	**TNAP2**	Q03169	Tumor necrosis factor alpha-induced protein 2	2.62938	
63	**MLKL**	Q8NB16	Mixed lineage kinase domain-like protein	3.39344	
64	**E7ETK0**	E7ETK0	40S ribosomal protein S24	3.59647	
65	**DHX36**	Q9H2U1	ATP-dependent DNA/RNA helicase DHX36	3.02912	
66	**TBCC**	Q15814	Tubulin-specific chaperone C	2.65875	
67	**RL6**	Q02878	60S ribosomal protein L6	3.71789	
68	**TOM34**	Q15785	Mitochondrial import receptor subunit TOM34	3.64649	
69	**HP1B3**	Q5SSJ5	Heterochromatin protein 1-binding protein 3	3.88668	
70	**C19L1**	Q69YN2	CWF19-like protein 1	2.65963	
71	**RS17**	P08708	40S ribosomal protein S17	3.74528	
72	**E7ESA6**	E7ESA6	Focal adhesion kinase 1	2.64713	
73	**AP3M1**	Q9Y2T2	AP-3 complex subunit mu-1	2.75333	
74	**M0QXD6**	M0QXD6	General transcription factor IIF subunit 1	2.56235	
75	**WNT5A**	P41221	Protein Wnt-5a	2.59575	
76	**GMDS**	O60547	GDP-mannose 4,6 dehydratase	3.08654	
77	**I3L0X5**	I3L0X5	Sperm-associated antigen 7	3.3113	
78	**E9PH64**	E9PH64	NADH dehydrogenase [ubiquinone] 1 beta subcomplex subunit 9	2.95705	
79	**MIPEP**	Q99797	Mitochondrial intermediate peptidase	2.7807	
80	**A0A087WZR9**	A0A087WZR9	Pyrroline-5-carboxylate reductase	5.26058	
81	**CHIP**	Q9UNE7	E3 ubiquitin-protein ligase CHIP	2.93915	
82	**TBCE**	Q15813	Tubulin-specific chaperone E	2.57258	
83	**SAAL1**	Q96ER3	Protein SAAL1	2.84296	
84	**SYRC**	P54136	Arginine--tRNA ligase, cytoplasmic		3.68764
85	**HTAI2**	Q9BUP3	Oxidoreductase HTATIP2		2.52883
86	**GRP75**	P38646	Stress-70 protein, mitochondrial		3.35915
87	**PSB3**	P49720	Proteasome subunit beta type-3		2.82347
88	**GSTP1**	P09211	Glutathione S-transferase P		3.25767
89	**RAB7A**	P51149	Ras-related protein Rab-7a		2.95429

90	TCPG	P49368	T-complex protein 1 subunit gamma	3.5822
91	RAB9A	P51151	Ras-related protein Rab-9A	2.60002
92	PHB	P35232	Prohibitin	2.76765
93	RS27A	P62979	Ubiquitin-40S ribosomal protein S27a	3.44529
94	E7EQR4	E7EQR4	Ezrin	2.70739
95	RIR1	P23921	Ribonucleoside-diphosphate reductase large subunit	2.55098
96	J3QQT2	J3QQT2	60S ribosomal protein L17	2.99247
97	SRP68	Q9UHB9	Signal recognition particle subunit SRP68	3.01785
98	SMD3	P62318	Small nuclear ribonucleoprotein Sm D3	3.12363
99	SYQ	P47897	Glutamine--tRNA ligase	2.65577
100	H3BQI1	H3BQI1	Dynein light chain roadblock-type 2	2.72107
101	MCM5	P33992	DNA replication licensing factor MCM5	2.94611
102	Q8WVC2	Q8WVC2	40S ribosomal protein S21	2.76124
103	NDUS3	O75489	NADH dehydrogenase [ubiquinone] iron-sulfur protein 3, mitochondrial	3.63337
104	RL7	P18124	60S ribosomal protein L7	2.73981
105	C9JA28	C9JA28	Translocon-associated protein subunit gamma	3.0665
106	S4R3E9	S4R3E9	NEDD8-MDP1 readthrough	3.09505
107	VIME	P08670	Vimentin	3.24861
108	I3L504	I3L504	Eukaryotic translation initiation factor 5A-1	2.65408
109	TM109	Q9BVC6	Transmembrane protein 109	3.52926
110	DRG1	Q9Y295	Developmentally-regulated GTP-binding protein 1	3.14037
111	SMD1	P62314	Small nuclear ribonucleoprotein Sm D1	3.54617
112	H0YEN5	H0YEN5	40S ribosomal protein S2	3.76272
113	MCM3	P25205-2	DNA replication licensing factor MCM3	3.01339
114	ARF6	P62330	ADP-ribosylation factor 6	3.18205
115	RL7A	P62424	60S ribosomal protein L7a	3.56152
116	BZW2	Q9Y6E2	Basic leucine zipper and W2 domain-containing protein 2	4.20893
117	A0A0A0MTN0	A0A0A0MTN0	Cullin-2	2.85465
118	THOP1	P52888	Thimet oligopeptidase	3.49956
119	J3QRI7	J3QRI7	60S ribosomal protein L26	5.03306
120	ATPO	P48047	ATP synthase subunit O, mitochondrial	2.73623
121	LSM3	P62310	U6 snRNA-associated Sm-like protein LSm3	3.12073
122	SRP14	P37108	Signal recognition particle 14 kDa protein	2.97832
123	NCBP2	P52298	Nuclear cap-binding protein subunit 2	2.55743
124	RAN	P62826	GTP-binding nuclear protein Ran	2.58877
125	J3KQ48	J3KQ48	Peptidyl-tRNA hydrolase 2, mitochondrial	4.00593

126	TXD17	Q9BRA2	Thioredoxin domain-containing protein 17	3.25391
127	MGST1	P10620	Microsomal glutathione S-transferase 1	2.82746
128	SNX1	Q13596	Sorting nexin-1	2.5274
129	HNRPF	P52597	Heterogeneous nuclear ribonucleoprotein F	2.89938
130	RL13	P26373	60S ribosomal protein L13	3.9096
131	A0A087WUD3	A0A087WUD3	Oligosaccharyltransferase complex subunit OSTC	2.65514
132	RS28	P62857	40S ribosomal protein S28	5.78257
133	J3KTA4	J3KTA4	Probable ATP-dependent RNA helicase DDX5	4.43944
134	DDX6	P26196	Probable ATP-dependent RNA helicase DDX6	2.70974
135	ZC3HF	Q8WU90	Zinc finger CCCH domain-containing protein 15	2.64774
136	RM21	Q7Z2W9	39S ribosomal protein L21, mitochondrial	3.18026
137	ULA1	Q13564	NEDD8-activating enzyme E1 regulatory subunit	3.12154
138	RL11	P62913	60S ribosomal protein L11	4.33191
139	RS11	P62280	40S ribosomal protein S11	3.57281
140	RS8	P62241	40S ribosomal protein S8	4.43922
141	RL21	P46778	60S ribosomal protein L21	3.30215
142	A0A087X0X3	A0A087X0X3	Heterogeneous nuclear ribonucleoprotein M	2.92309
143	RS23	P62266	40S ribosomal protein S23	4.45084
144	ATD3A	Q9NVI7-2	ATPase family AAA domain-containing protein 3A	2.65768
145	RL23	P62829	60S ribosomal protein L23	4.39359
146	2A5D	Q14738-3	Serine/threonine-protein phosphatase 2A 56 kDa regulatory subunit delta isoform	3.24429
147	RS13	P62277	40S ribosomal protein S13	5.13601
148	G5E9Q6	G5E9Q6	Profilin	3.91099
149	RL12	P30050	60S ribosomal protein L12	4.5988
150	XPO5	Q9HAV4	Exportin-5	2.81304
151	DDX23	Q9BUQ8	Probable ATP-dependent RNA helicase DDX23	3.09004
152	RL27	P61353	60S ribosomal protein L27	5.56322
153	E5RI99	E5RI99	60S ribosomal protein L30	4.79294
154	SHIP2	O15357	Phosphatidylinositol 3,4,5-trisphosphate 5-phosphatase 2	2.93228
155	PRP6	O94906	Pre-mRNA-processing factor 6	4.71226
156	SYDM	Q6PI48	Aspartate--tRNA ligase, mitochondrial	4.19994
157	KIF2A	O00139-2	Kinesin-like protein KIF2A	2.70644
158	J3QR09	J3QR09	Ribosomal protein L19	2.74218
159	OCAD2	Q56VL3	OCIA domain-containing protein 2	4.79686
160	J3QSV6	J3QSV6	Ribosomal L1 domain-containing protein 1	4.28401
161	RL38	P63173	60S ribosomal protein L38	2.93381
162	A0A087X2D0	A0A087X2D0	Serine/arginine-rich-splicing factor 3	4.36894

163	GPI8	Q92643	GPI-anchor transamidase		2.86273
164	NU160	Q12769	Nuclear pore complex protein Nup160		2.76185
165	SMYD3	Q9H7B4	Histone-lysine N-methyltransferase SMYD3		3.32229
166	YMEL1	Q96TA2	ATP-dependent zinc metallo-protease YME1L1		2.84011
167	PDCD4	Q53EL6	Programmed cell death protein 4		3.59557
168	TMED1	Q13445	Transmembrane emp24 domain-containing protein 1		2.72926
169	RCN2	Q14257	Reticulocalbin-2		3.22922
170	R39L5	Q59GN2	Putative 60S ribosomal protein L39-like 5		5.63947
171	SRP09	P49458	Signal recognition particle 9 kDa protein		4.54385
172	TSR1	Q2NL82	Pre-rRNA-processing protein TSR1 homolog		2.96676
173	DYR	P00374	Dihydrofolate reductase		4.23334
174	DHX8	Q14562	ATP-dependent RNA helicase DHX8		2.56375
175	RIPK1	Q13546	Receptor-interacting serine/threonine-protein kinase 1		3.70312
			Common in both WT- and P301L- tau expressing cells		
176	F8WCF6	F8WCF6	Actin-related protein 2/3 complex subunit 4	3.33715	4.14691
177	RACK1	P63244	Receptor of activated protein C kinase 1	3.39824	4.51018
178	SYDC	P14868	Aspartate--tRNA ligase, cytoplasmic	2.74959	2.57307
179	EIF3K	Q9UBQ5	Eukaryotic translation initiation factor 3 subunit K	2.81828	4.58194
180	H9KV45	H9KV45	Ubiquitin-conjugating enzyme E2 D3	2.67823	5.42209
181	Q5JR08	Q5JR08	Rho-related GTP-binding protein RhoC	3.44748	3.81268
182	PROF1	P07737	Profilin-1	3.07127	4.69295
183	FPPS	P14324	Farnesyl pyrophosphate synthase	3.80014	3.24431
184	Q5VV89	Q5VV89	Microsomal glutathione S-transferase 3	3.6279	4.42465
185	SYK	Q15046	Lysine--tRNA ligase	3.9661	3.1659
186	PUR2	P22102	Trifunctional purine biosynthetic protein adenosine-3 [Includes: Phosphoribosylamine--glycine ligase	2.81901	2.93211
187	SND1	Q7KZF4	Staphylococcal nuclease domain-containing protein 1	3.019	3.608
188	C9JZR2	C9JZR2	Catenin delta-1	2.6207	2.99525
189	CAN2	P17655	Calpain-2 catalytic subunit	3.42537	3.11891
190	AAAS	Q9NRG9	Aladin	4.28764	4.41039
191	SYLC	Q9P2J5	Leucine--tRNA ligase, cytoplasmic	3.39758	3.6911
192	RNZ2	Q9BQ52	Zinc phosphodiesterase ELAC protein 2	3.34161	3.22
193	TIF1B	Q13263	Transcription intermediary factor 1-beta	3.23108	3.44094
194	EIF3B	P55884	Eukaryotic translation initiation	4.32237	4.51386

			factor 3 subunit B		
195	**PPIA**	P62937	Peptidyl-prolyl cis-trans iso-merase A	3.8054	4.17
196	**H0YAK1**	H0YAK1	G-rich sequence factor 1	3.46677	2.53854
197	**EIF3G**	O75821	Eukaryotic translation initiation factor 3 subunit G	2.72457	3.04545
198	**Q32Q12**	Q32Q12	Nucleoside diphosphate kinase	3.00854	3.04007
199	**EFTU**	P49411	Elongation factor Tu, mitochondrial	3.64175	3.42478
200	**RS3**	P23396	40S ribosomal protein S3	2.61813	3.38603
201	**MTREX**	P42285	Exosome RNA helicase MTR4	2.57741	2.61761
202	**RS26**	P62854	40S ribosomal protein S26	2.66498	3.15958
203	**RS16**	P62249	40S ribosomal protein S16	2.87042	5.66172
204	**A0A087WZT3**	A0A087WZT3	BolA-like protein 2	2.62918	3.9663
205	**LANC1**	O43813	Glutathione S-transferase LANCL1	4.28469	3.09074
206	**CND3**	Q9BPX3	Condensin complex subunit 3	2.9969	2.66232
207	**AIMP2**	Q13155	Aminoacyl tRNA synthase complex-interacting multifunctional protein 2	3.83739	3.0309
208	**G3V325**	G3V325	ATP5MF-PTCD1 readthrough	4.35305	4.77429
209	**TGM2**	P21980	Protein-glutamine gamma-glutamyltransferase 2	3.47902	2.55024
210	**RS18**	P62269	40S ribosomal protein S18	3.20805	4.17992
211	**RS14**	P62263	40S ribosomal protein S14	2.67833	4.25944
212	**C9JXB8**	C9JXB8	60S ribosomal protein L24	2.98512	4.69503
213	**MGST2**	Q99735	Microsomal glutathione S-transferase 2	2.68604	3.32233
214	**RL13A**	P40429	60S ribosomal protein L13a	2.87139	4.09016
215	**SNR40**	Q96DI7	U5 small nuclear ribonucleo-protein 40 kDa protein	2.83902	2.94416
216	**SRP54**	P61011	Signal recognition particle 54 kDa protein	3.37647	3.10435
217	**CUL4A**	Q13619	Cullin-4A	5.00644	2.58406
218	**UCK2**	Q9BZX2	Uridine-cytidine kinase 2	2.99612	2.78459
219	**K7EP65**	K7EP65	60S ribosomal protein L22	3.68703	6.73393
220	**RS19**	P39019	40S ribosomal protein S19	3.54415	7.35693
221	**TSYL1**	Q9H0U9	Testis-specific Y-encoded-like protein 1	2.66813	3.09096
222	**E9PJD9**	E9PJD9	60S ribosomal protein L27a	3.04224	5.20122
223	**VPS45**	Q9NRW7	Vacuolar protein sorting-associated protein 45	2.73297	4.2448
224	**M0R3D6**	M0R3D6	60S ribosomal protein L18a	3.29172	4.15871
225	**RS6**	P62753	40S ribosomal protein S6	2.85538	3.91982
226	**H7C2W9**	H7C2W9	60S ribosomal protein L31	2.97225	6.33005
227	**NUCL**	P19338	Nucleolin	3.49248	4.59942
228	**A0A0D9SG12**	A0A0D9SG12	ATP-dependent RNA helicase DDX3X	2.69323	4.56215
229	**CDC73**	Q6P1J9	Parafibromin	3.60285	2.99202
230	**APT**	P07741	Adenine phosphoribosyltransferase	3.34193	2.61598
231	**E7ENU7**	E7ENU7	Ribosomal protein L15	3.97322	2.8185
232	**G3BP1**	Q13283	Ras GTPase-activating	3.04396	3.61278

			protein-binding protein 1		
233	**K7ERT8**	K7ERT8	60S ribosomal protein L23a	4.22784	5.36401
234	**J3QQ67**	J3QQ67	60S ribosomal protein L18	2.66443	2.95231
235	**XRCC6**	P12956	X-ray repair cross-complementing protein 6	3.48879	3.75823
236	**M0R0R2**	M0R0R2	40S ribosomal protein S5	4.26617	5.14718
237	**RFC5**	P40937	Replication factor C subunit 5	4.36703	3.35394
238	**F8W727**	F8W727	60S ribosomal protein L32	3.74762	5.39745
239	**MD2L1**	Q13257	Mitotic spindle assembly checkpoint protein MAD2A	2.92545	2.98028
240	**D6RAN4**	D6RAN4	60S ribosomal protein L9	3.31616	4.19725
241	**RL4**	P36578	60S ribosomal protein L4	2.85467	3.7287
242	**RL36**	Q9Y3U8	60S ribosomal protein L36	4.53022	5.5977
243	**SSBP**	Q04837	Single-stranded DNA-binding protein, mitochondrial	5.65442	4.22935
244	**RS10**	P46783	40S ribosomal protein S10	3.25188	3.02068
245	**C9JZI1**	C9JZI1	Replication factor C subunit 4	2.62627	3.47232
246	**XRCC5**	P13010	X-ray repair cross-complementing protein 5	2.60763	3.48406
247	**AMPN**	P15144	Aminopeptidase N	3.75088	3.23543
248	**GTF2I**	P78347	General transcription factor II-I	2.60151	3.52247
249	**RS7**	P62081	40S ribosomal protein S7	3.97012	3.42622
250	**C9JQV0**	C9JQV0	Uncharacterized protein C7orf50	2.86837	2.93894
251	**ACOX3**	O15254	Peroxisomal acyl-coenzyme A oxidase 3	2.7579	4.58583

Table 18: Top 5 networks identified from IPA analysis in WT-tau expressing cells

ID	Associated network functions	Score
1	RNA damage and repair, protein synthesis, cancer	61
2	DNA replication, recombination, and repair, cellular assembly and organization, cell morphology	41
3	Cell cycle, cellular function and maintenance, molecular transport	28
4	Cellular movement, haematological system development and function, immune cell trafficking	21
5	Organismal injury and abnormalities, cellular movement, skeletal and muscular system development and function	17

Table 19: Top 5 networks identified from IPA analysis in P301L-tau expressing cells

ID	Associated network functions	Score
1	RNA damage and repair, protein synthesis, cancer	55
2	Cellular assembly and organization, cellular compromise, cellular function and maintenance	52
3	Nucleic acid metabolism, small molecule biochemistry, digestive system development and function	27
4	Cellular compromise, developmental disorder, endocrine system disorders	22
5	Developmental disorder, endocrine system disorders, organ morphology	11